ETHICAL ISSUES IN THE 21ST CENTURY

DRONE WARFARE

ETHICAL EXPLORATIONS

ETHICAL ISSUES IN THE 21ST CENTURY

Additional books in this series can be found on Nova's website
under the Series tab.

Additional e-books in this series can be found on Nova's website
under the e-book tab.

ETHICAL ISSUES IN THE 21ST CENTURY

DRONE WARFARE

ETHICAL EXPLORATIONS

MICHELLE HOLLOWAY
EDITOR

New York

NOTICE TO THE READER

Library of Congress Cataloging-in-Publication Data

ISBN: 978-1-63485-103-9

Published by Nova Science Publishers, Inc. † *New York*

CONTENTS

PREFACE

With the increased use of armed drones has come closer scrutiny of the legal and ethical dimensions. In the first chapter of this book, the author makes an assessment of the lethal use of drone technologies, measured in terms of their legality, morality, and overall effectiveness. Armed unmanned aerial vehicles—combat drones—have fundamentally altered the ways the United States conducts military operations aimed at countering insurgent and terrorist organizations. Drones may reduce risks to human soldiers but the question arises as to whether they permit the initiation or escalation of conflict by promoting civic disengagement. The authors of the second chapter offer an analysis of the dimensions surrounding this argument.

In: Drone Warfare: Ethical Explorations   ISBN: 978-1-63485-103-9
Editor: Michelle Holloway   © 2016 Nova Science Publishers, Inc.

Chapter 1

LETHAL AND LEGAL?
THE ETHICS OF DRONE STRIKES[*]

Shima D. Keene

FOREWORD

The use of drones, or unmanned aerial vehicles, has increased exponentially in the last 10 years, and this trend is likely to continue for the foreseeable future. But with this increased use has come increased controversy, in particular closer scrutiny of the legal and ethical dimensions of the use of armed drones.

In this monograph, British academic and practitioner Dr. Shima Keene provides a comprehensive assessment of the lethal use of drone technologies, measured in terms of their legality, morality, and overall effectiveness in the fight against terrorism and counterinsurgency operations. Dr. Keene is a subject matter expert in the field of asymmetric warfare and counterterrorism, and a former Director of the Security Technology Short Course at the Defence Academy of the United Kingdom, where she conducted research into both technical and ethical aspects of the deployment of UAV technologies. In her monograph, she explores the legal and ethical bases for lethal use of drones, both from a U.S. and an international perspective. Dr. Keene also highlights knowledge gaps that must be filled in order to be able to make an accurate

---

[*] This is an edited, reformatted and augmented version of a monograph issued by the Strategic Studies Institute, December, 2015.

assessment of the success or failure of operations where drones have been deployed, and argues that greater transparency is needed to obtain broad public support for their use.

The Strategic Studies Institute considers that this monograph provides a useful assessment of the key issues relating to the legality, morality, and effectiveness of drone use, and is a valuable addition to the debate on how to plan and shape future U.S. operations involving unmanned and autonomous technologies.

DOUGLAS C. LOVELACE, JR.
Director
Strategic Studies Institute and U.S. Army War College Press

SUMMARY

With greatly increased lethal use of unmanned aerial vehicles (UAVs) comes greater scrutiny and controversy. This monograph lays out the ethical and legal landscape in which drone killings take place and makes key recommendations not only for ensuring legality and a sound moral basis for operations, but also for ensuring those operations are effective.

While supporters claim that drone warfare is not only legal but ethical and wise, others have suggested that drones are prohibited weapons under International Humanitarian Law (IHL) because they cause, or have the effect of causing indiscriminate killings of civilians, such as those in the vicinity of a targeted person. The main legal justification made by the Barack Obama administration for the use of armed drones is self-defense. However, there is ambiguity as to whether this argument can justify a number of recent attacks by the United States. In order to determine the legality of armed drone strikes, other factors such as sovereignty, proportionality, the legitimacy of individual targets, and the methods used for the selection of targets must also be considered.

The ethical landscape is also ambiguous. One justification is the reduced amount of collateral damage possible with drones relative to other forms of strike. Real-time eyes on target allow last-minute decisions and monitoring for unintended victims, and precise tracking of the target through multiple systems allows further refinements of proportionality. But this is of little benefit if the definition of "targets" is itself flawed and encompasses noncombatants and unconnected civilians.

This monograph also provides a number of specific recommendations intended to ensure that the benefits of drone warfare are weighed against medium- and long-term second order effects, so as to measure whether targeted killings are serving their intended purpose of countering terrorism rather than encouraging and fueling it.

INTRODUCTION

No aspect of modern warfare is as controversial as the use of armed drones. Everything about drone technology is contested: its novelty, legality, morality, utility, and future development. Even the choice of what to call such systems is value-laden.[1]

Professor Sir David Omand GCB Former United Kingdom Security and Intelligence Coordinator

In the years since the attacks of September 11, 2001 (9/11), the United States has developed the use of unmanned aircraft (UA)—drones—to locate, target, and eliminate individuals overseas considered to pose a threat to the United States. Although the program was initially kept secret, in recent years the U.S. Government has acknowledged that drones are used to target members of al-Qaeda and associated forces within theaters of conflict, as well as outside it. In the last 10 years, this use of drones has increased exponentially.[2]

The overall use of armed drones by the United States is reported to have grown by 1,200 percent between 2005 and 2013.[3] More drone strikes were authorized by the Barack Obama administration in 2009 than by that of George W. Bush during his entire time in office,[4] and by early-2012, the Pentagon was reported to have 7,500 drones under its control, representing approximately one-third of all U.S. military aircraft.[5] Furthermore, this trend is expected to continue, with many viewing unmanned vehicles as the future of warfare.[6]

Key rationales for the use of armed drones are that they are legal, effective, and ethical. According to White House spokesman James Carney speaking in February 2013, the current U.S. administration views strikes by drones to be "legal, ethical, and wise." However, the simple fact that examination of the ethics of drone strikes has been included in the 2014 Army Priorities for Strategic Analysis testifies to the moral ambiguity that persists surrounding their use as a tool of policy.

WHAT IS A DRONE?

The term "drone" refers to any UA controlled remotely by an operator on the ground. Historically, the term "unmanned aerial vehicles" (UAVs) was used, although this is now seen by many as misleading since there is a pilot, albeit on the ground as opposed to in the aircraft itself.[7] Consequently, the term "remotely piloted aircraft" (RPA) has become the preferred term for some as it provides a more accurate description as to what a drone is.

The term "drone" dates from the 1930s. It is believed to originate with the "Queen Bee" radio controlled aircraft,[8] the first returnable and reusable UAV developed in the United Kingdom (UK) and designed for use in air and naval gunnery practice.[9] Today, the desire to avoid use of the term "drone" by some governments outside of the United States reflects that the word has come to imply morally dubious "killing machines" responsible for the murder of innocent civilians. This is a direct result of media reporting on the U.S. weaponized drones program.

Media representation of armed drones generally has been negative, which is partly the reason why the British Ministry of Defence (MOD) has rejected the term in favor of RPAs. The search for an alternative term for "drone" is also representative of a desire to remind people of other uses for drones which are often forgotten. Even within a military context, drones are used in a nonkinetic as well as a kinetic capacity, including for surveillance and intelligence gathering. In addition, their use is not restricted to counterterrorism and counterinsurgency (COIN) operations, but extends to humanitarian peacekeeping and peace enforcement operations as well.

Drone technologies offer many benefits, including significant economic value and social benefits, and are increasingly being exploited by civil authorities responsible for safety, security, and policing.[10] Outside defense and security, the technology is also utilized by a number of industries for civilian applications including agriculture, media,[11] catering, private security, law enforcement, conservation, and environmental monitoring. Several of these domains—most notably the media—are also subject to ongoing debates on ethics and legality, focusing on privacy and the need for regulation. Recent examples include use by the Police Service of Northern Ireland during the G8 summit in 2013,[12] by German national railway company Deutsche Bahn to track graffiti artists on its property,[13] and by the Japanese restaurant chain Yo!Sushi, which recently introduced the "ITray,"[14] a custom built flying platter, in its Soho branch in London, UK.[15] Drones have also been used in South Africa for tracking poachers, resulting in a reduction in rhinoceros deaths.[16]

Further potential applications under consideration include the airborne distribution of pesticides, fire investigations in unsafe buildings, search and rescue missions in treacherous conditions, police searches for missing or wanted persons, traffic management, public order situations, and evidence gathering.[17]

The potential benefits of drone technologies for military and nonmilitary applications are considerable. The first benefit is cost. As the aircraft does not have to be built around a human, it can be of any size. The lack of a pilot and passengers also negates the need for support systems such as pressurized cabins, further reducing cost. The second benefit is the improved duration of flying times, which can be increased from hours to weeks. The third benefit is the ability to risk dangerous conditions such as extreme poor weather, which would be outside minima if there were a pilot on board. Unsurprisingly, therefore, the use of drone technology is expected to continue to increase across all applications.

For the advocates of drone warfare, the evolution of drones represents a natural advancement in technology which is regarded as both logical and welcome. Many also accept drones as an inevitability given that we live in an era of rapidly evolving technological advancement, coupled with a climate of economic austerity where there is a constant requirement to deliver more for less cost. This cost refers not only to financial cost, but also cost in lives. The latter criterion is also attractive in policy terms, because it reduces or eliminates the risk of friendly casualties. In casualty-averse Western societies, this can be a key determinant of the sustainability of military engagement.

At the same time, for others, the use of drones signals a dangerous decline in morality and accountability. The use of drones for targeted killings has been described as a step change in warfare, not only in terms of technological capabilities, but in terms of the ethical and legal frameworks that have governed the use of force for decades. Critics argue that the normalization of the use of drones represents a slippery slope that intrudes upon human rights, and increases the temptation to use force while diminishing the accountability of those engaged in such actions.

There are clearly a number of considerations regarding the armed use of drones. The purpose of this monograph is to explore the answers behind three key questions: First, is it legal? Second, is it ethical? Third, is it effective? Each will be examined in turn.

Is It Legal?

The main legal justification made by the Obama administration for the use of armed drones is self-defense. It is argued that following the 9/11 attacks in 2001, the United States is defending itself against enemies who are constantly contemplating and planning deadly attacks against it. Furthermore, according to State Department legal advisor Harold Hongju Koh in a public statement made in March 2010, the U.S. practice of targeted killing complies fully with all applicable law.[18] The U.S. position is that as part of the ongoing war on terror, the circumstances and the level of implied threat from terrorism is such that terrorists should be denied protection from International Humanitarian Law (IHL).

However, there is ongoing debate as to whether these actions are, in fact, legal. Warnings from critics such as Christof Heyns, the United Nations (UN) special rapporteur on extrajudicial killings, that the use of armed drones could constitute war crimes,[19] are clearly of concern, requiring further investigation.[20] Determining the legality of the use of armed drones to international standards is far from straightforward. First and foremost, there is no central legislative body or controlling authority for international law.[21] International treaties and state practices generally are considered to be the most authoritative sources of international law. But the breadth of factors that can be considered as sources of international law result in a profusion of authority, which can be problematic in establishing precedence. Furthermore, there is no obviously applicable international court to determine the legality or otherwise of drone strikes.

A further challenge is the ambiguous treatment of irregular warfare in international law. Under IHL, wars are recognized as armed conflicts fought between two or more states or high contracting parties.[22] Where this is not the case, existing international law only recognizes the existence of armed conflict in the context of civil or internal wars, and not between states and substate entities, namely irregular forces. In other words, international law is not fully equipped to tackle the 21st century security environment, characterized by asymmetric warfare, where nonstate actors typically take center stage.

This in many ways is central to the debate on the war on terror, and the role of armed drones within it. This war is not conventional warfare in that the enemy is not a state, but also because combatting the threat has both military and law enforcement elements. This is particularly true with the use of armed drones, which are used extensively by the Central Intelligence Agency (CIA) for counterterrorism purposes. Regardless of the extent to which the CIA may be considered a law enforcement agency, the tactics adopted have a closer resemblance to those used by law enforcement to tackle criminal gangs than to those of the military in conventional warfare.

However, even the distinction between law enforcement and military activities is often unclear as a result of the close cooperation between the CIA and the military. This is particularly true in counterterrorism operations, where a shift from "boots on the ground" to that of more discreet and deniable attacks has been observed.[23] According to Admiral William McRaven, former commander of Joint Special Operations Command and head of U.S. Special Operations Command:

> American military and intelligence operatives are virtually indistinguishable from each other as they carry out classified operations in the Middle East and Central Asia. [24]

This results in an ambiguity between the applicability of the law of armed conflict (LoAC) and the criminal justice approach. While the legality of drone attacks conducted by the military can be assessed under LoAC, when the attacks are carried out by law enforcement the assessment of legality becomes more complex. As a result, a number of existing legal sources and guidance must be used in attempting to determine the legality of armed drone attacks. A good starting point is to make an assessment on the legality of the military use of armed drones, which is governed by the LoAC. As such, a brief discussion of the legal concepts of *jus ad bellum* and *jus in bello*, which underpin the law of armed conflict, is essential.

Jus Ad Bellum versus Jus in Bello

Under international law, two aspects of warfare are considered. The first relates to **why** you are fighting, referring to whether the reasons for fighting can be justified. The second consideration is **how** you fight, which examines whether warfighting is conducted in a legitimate manner. In terms of the first, the reasons as to "why" you are fighting falls under the doctrine of "just war" theory, governed by *jus (or ius) ad bellum,* which is the title given to the branch of law that defines the legitimate reasons a state may engage in war. This is also referred to as the law of international armed conflict. It focuses on certain criteria, including; authority, just cause, right, and intention, which must be considered prior to engaging in war to determine whether it is a just war.

Jus in bello, on the other hand, is the set of laws that govern the way in which warfare is conducted and come into effect once a war has begun.[25] Its purpose is to regulate how wars are fought, without prejudice to the reasons as to how or why they had begun, and whether or not the cause upheld by either party is just. *Jus in bello* is also referred to as international humanitarian law, and the two terms are used interchangeably by the International Committee of the Red Cross (ICRC) as well as scholars who wish to emphasize their goal of mitigating the excesses of war and protecting civilians and other noncombatants. As its purpose is to limit the suffering in war by protecting and assisting victims as much as possible, IHL regulates only those aspects of conflict which are of humani- tarian concern.

The theoretical separation between *jus ad bellum* and *jus in bello* is also important, and should be noted.

In theory, it is possible to be engaged in an "unjust" war while adhering to the laws of armed conflict, or to be breaking the laws of armed conflict while fighting a just war. In addition, it is also often difficult to determine which state is guilty of violating the UN Charter that underpins *jus ad bellum*.[26] The application of humanitarian law does not involve the denunciation of guilty parties as that would be bound to arouse controversy and paralyze implementation of the law, since each adversary would claim to be a victim of aggression. This is why the two branches of law are, and should remain, completely independent of one another with the purpose of guaranteeing the application of *jus in bello,* irrespective of whether the war meets the *jus ad bellum* criteria.[27] At the same time, this necessity for separation also results in the inevitable tension between the two bodies of law. Each has its own historical origins and has developed in response to different values and objectives. Also, the fact that most of the principles of *jus in bello* predate the prohibition of the indiscriminate use of force[28] has led some to conclude that "modern" *jus ad bellum* has rendered IHL superfluous.

The distinction and the need to satisfy criteria set out by *jus in bello* and *jus ad bellum* has also been challenged by the view that, in some cases, a situation of self-defense may be so extreme and the threat to the survival of the state so great, that violations of *jus in bello* may be justified.[29] However, the humanitarian argument is that victims on both sides of a conflict are equally worthy of protection. As such, the need for separation of the two bodies of law based on humanitarian grounds is equally convincing.[30] While there is no clear consensus as to which should take priority, there are additional considerations that should be examined in determining the legality of armed drones.

The first factor to consider is the legitimacy of targets and targeting methods, referring to the process used to identify, prioritize and select or eliminate targets which are considered to be of operational and strategic value. Targets can be mobile targets such as individuals or groups of individuals, or stationary targets such as lines of communications or hardened facilities. Although damage to stationary targets can also have considerable negative impact on local populations, which must be taken into consideration when assessing proportionality, this monograph will focus on the more emotive question of directly targeting humans.

Determining the Legitimacy of a Target

For an armed attack to be deemed legitimate, the individual targeted must also be a legitimate target in the eyes of the law. Legal guidance is available through UN and other rulings as to what constitutes a valid target. However, the reality is not always clear cut. The media reported that the Joint Integrated Prioritized Target List produced by the U.S. Department of Defense (DoD) in 2009[31] contained just over 400 names of individuals described as "known terrorists," who are deemed to be legitimate targets.[32] However, many argue that the definition of "terrorist" used is far too broad to be legally defensible for targeting decisions.[33]

The inclusion of Taliban financiers on the so-called "kill list" is a case in point. Although terrorist financiers may be classified as terrorists, there is debate as to whether they can be deemed a legitimate target for a kill list under IHL. According to ICRC guidance, IHL stipulates that the decisive criterion as to whether an individual is a member of an organized armed group, terrorist or otherwise, is to prove a continuous combat function. Furthermore, any individual falling outside this category would be classified as a civilian. In other words, unless a Taliban financier can be proved to be a combatant on a continual basis, it would be unlawful to target him under IHL.

The next factor that must be taken into consideration is distinction, referring to the ability to differentiate between an individual who is a terrorist and who is not. This can be particularly challenging when attempting to isolate the target from his or her family. Under IHL, if family members fall victim, it is not considered to be a legal kill, unless it can be proved that the family members were also part of a targeted organization and causing real harm. Unfortunately, reporting on U.S. drone strikes to date gives the impression that this criterion has been ignored. If this is true, greater attention to legality is necessary. If it is false, greater attention to media and perception management is needed.

A further point of legal debate relates to the circumstances in which a terrorist is killed, in particular with reference to "rescuer attacks" or "follow-up strikes." One example is the killing of senior al-Qaeda leader Abu Yahya Al-Libi on June 4, 2012. Following an initial drone strike which killed five people and injured four others, a group of 12 people, including local residents, came to the assistance of the victims. Al-Libi was reported to have been overseeing the rescue efforts and was killed in the second strike, along with between 9 and 15 other people, including six local tribesmen. In other words, six civilians were killed working in a humanitarian capacity alongside a group of al-Qaeda operatives under a senior al-Qaeda official.

The follow-up strike has been described as a potential war crime both because it constituted an attack on civilian rescuers, and also because al-Libi may not have been directly participating in hostilities at the time of the strike. However, the question as to whether al-Libi was directly participating in hostilities and therefore deemed as a legitimate target does not depend on what he was doing at the time he was killed.[34] What really matters in terms of determining the legality of the attack is whether the attack was carried out as part of an actual armed conflict (such as the non-international armed conflict [NIAC][35] in Afghanistan) and his role in that conflict.[36] There is an additional argument that even suspected terrorists should have the right to surrender and defend themselves in court. The right to surrender and defense in court is equally applicable in relation to selecting targeted killing by drones over the option of capture. This argument put aside, to qualify as a target of a kill list, the individual must be considered to pose a direct threat to the United States for the act to be deemed legitimate.

Furthermore, a kill list is not the only way the United States targets individuals using drones. A significant proportion of the individuals killed in drone strikes are not, even by the U.S. Government's account, militant leaders and thus are unlikely to be on a kill list. As such, the method by which a target is selected becomes highly relevant. In other words, how someone is killed is considered to be as important as who is killed, as this will determine whether a killing can be deemed a targeted killing or an assassination.[37] The significance of the differentiation from a legal standpoint is that targeted killings are legitimate whereas assassinations are not viewed as legitimate under either domestic or international law.[38] Furthermore, targeted killings must be executed using conventional military means only.[39] When treachery is deemed to exist, where targets have been misled, and deception is used, a targeted killing cannot be justified[40] and may be considered to be an assassination.

Sovereignty

The geographical location where the drone attacks are taking place, and consideration of sovereignty, together with possible violation of the territorial integrity of the countries where targeted killing takes place, are also key factors in determining the legality of a drone strike. As a threshold matter, the *jus ad bellum* inquiry depends on whether the "host state" has consented to the drone strike. If there is consent, there is no infringement on sovereignty. Publicly available literature suggest that many of the states where the drone attacks are taking place, such as Iraq, have provided consent to the United States, thus making the attacks legal from the perspective of sovereignty.

However, the situation in jurisdictions such as Pakistan is less clear. On the one hand, the United States has claimed that it is acting with the consent of Pakistan.[41] At the same time, Pakistan has publicly denied this.[42] According to the United States, this is because the Pakistani government believes that the decision to give consent would be unpopular with the Pakistani people.[43] In other words, they are covertly supportive of U.S. action, but for political reasons feel that they must be seen to be opposing it. As such, a definitive answer to this factual question is impossible without access to confidential material. As a result of the lack of explicit consent, alternative justifications to provide a legal basis for the continued U.S. drone strikes in Pakistan become necessary. Here, the United States has turned to Article 51 of the UN Charter, which preserves each state's "inherent right of individual or collective self-defense if an armed attack occurs."

The Law of Self-Defense

In June 2014, the New York City Bar Committee on International Law published a report entitled "The Legality under International Law of Targeted Killings by Drones launched by the United States."[44] In determining the applicability of the law of self-defense, the Committee concluded that the right to self-defense was available against nonstate actors provided there is an actual or threatened "armed attack" by the non-state actor, and that acts of violence by nonstate actors can rise to the level of an "armed attack" within the meaning of Article 51, if they are of sufficient scale and effect. Although not all acts of terrorism justify the use of armed force, as opposed to a law enforcement response, a single act of terrorism may constitute an "armed attack" if it is of sufficient intensity.

The committee found that the 9/11 attacks in 2001 constituted an "armed attack" by al-Qaeda on the U.S., giving rise to a right of armed self-defense against alQaeda pursuant to Article 51 of the UN Charter. Consequently, the invasion of Afghanistan was deemed to be a legitimate exercise of force in self-defense. However, the committee concluded that the 9/11 attacks alone no longer supply a self-defense legal basis for additional measures taken against al-Qaeda. For the continued use of force to be justified on the basis of self-defense, it must be defensible through current "armed attacks," and therefore the use of force world-wide against organizations that are not al-Qaeda core, including any alleged "affiliates" of al-Qaeda, cannot be justified as a *jus ad bellum* matter by the attacks of 9/11 alone. However, under some circumstances, the accumulation of smaller acts of violence committed by a nonstate actor may constitute an "armed attack," provided that the use of force in self-defense is constrained by the principles of necessity and proportionality. As such, another factor in determining the legality of a targeted killing is that the action must be deemed to be proportionate.

Proportionality

Proportionality is a fundamental principle of *jus in bello*,[45] which is codified in the First Additional Protocol to the 1949 Geneva Conventions. The principle prohibits:

> an attack which may be expected to cause incidental loss of civilian life, injury to civilians, damage to civilian objects, or a combination thereof, which would be excessive in relation to the concrete and direct military advantage anticipated.[46]

The killing of civilians in itself does not make an attack unlawful. However, in determining proportionality, anticipated collateral damage must be taken into consideration. This is conducted as an *ex ante* analysis, as opposed to an *ex post* measure of the actual outcome. In addition, an assessment of the expected military advantage that the attack will confer on the attacker is needed. However, attempting to determine the direct military advantage needed to justify the attack is not straightforward as there is debate with respect to whether the specific benefit of the attack should be viewed in isolation[47] or in light of the attack's role in the overarching military objective.[48] Finally, the anticipated collateral damage must be weighed against the military benefit to ensure that the former is not excessive compared to the latter.[49] The key challenge here is the lack of facts, either on the projected collateral damage or on the expected military benefit, let alone how to balance the two, making the analysis of the proportionality of individual drones strikes impossible.[50]

Also, for a targeted killing to be legitimate, the target must represent a direct and imminent threat to the United States. However, some analysts believe that, today, al-Qaeda constitutes more of an ideological influence than a genuine direct threat.[51] According to Brian Jenkins of the Rand Corporation:

> The architects of 9/11 have been captured or killed. Al Qaeda's founder and titular leader is dead. Its remaining leadership has been decimated. The group's wanton slaughter of Muslims has alienated much of its potential constituency. Cooperation among security services and law enforcement organizations world-wide has made its operating environment more hostile. Al Qaeda has not been able to carry out a significant terrorist operation in the West since 2005, although, as demonstrated on the tenth anniversary of 9/11, it is still capable of mounting plausible, worrisome threats.[52]

The general consensus is that in 2015, 9/11 can no longer be used as a justification for self-defense. However, as a threat from affiliate Islamic militant groups clearly persists, the legal position with respect to groups such as the Islamic State (IS) remains unclear. This is particularly true as regards the use of armed drones. As the *International Review of the Red Cross* observed:

> Armed drones pose a major threat to the general prohibition on the inter-state use of force and to respect for human rights. On the battlefield, in a situation of armed conflict, the use of armed drones may be able to satisfy the fundamental international humanitarian law rules of distinction and proportionality. Away from the battlefield, the use of drone strikes will often amount to a violation of fundamental human rights. Greater clarity on the applicable legal regime along with restraints to prevent the further proliferation of drone technology are urgently needed.[53]

What is also clearly much needed is greater transparency and accountability in order that an informed debate can take place in the public domain in relation to the legality of armed drone attacks. The subject of legality is related to ethics and morality which form the basis of legislation. The subject of ethics relating to drone attacks will be examined next.

Is It Ethical?

Advocates of drone warfare argue that the use of drones is ethical, especially when the alternative is the use of airstrikes—a blunt instrument which will result in greater collateral damage, as well as risking the lives of U.S. military personnel. Some go even further to argue that the United States is not only legally entitled to use drones, but is morally obliged to do so on the grounds of safety and accuracy. According to Bradley J. Strawser,[54] an assistant professor in the defense analysis department at the U.S. Naval Postgraduate School and a research associate with Oxford's Institute for Ethics, Law, and Armed Conflict:

> . . . It's all upside. There's no downside. Both ethically and normatively, there's a tremendous value. You're not risking the pilot. The pilot is safe. And all the empirical evidence shows that drones tend to be more accurate. We need to shift the burden of the argument to the other side. Why not do this? The positive reasons are overwhelming at this point. This is the future of all air warfare. At least for the U.S. [55]

Others do not share this view. For example, former President Jimmy Carter has expressed his unease with reference to the White House kill lists claiming that: ". . . the U.S. can no longer speak with moral authority on human rights."[56]

Strawser's viewpoint may be correct to the extent of the safety of the drone pilot, but the subject of accuracy is disputed,[57] especially when dynamic targeting methods are adopted. However, as a starting point on the debate on ethics of drone strikes, the ethical argument can only be made if the correct targets have been identified and the killing has been carried out with minimal collateral damage.

Are the Drones Targeting the Right People?

The previous section on legality has already highlighted a number of targeting legitimacy challenges, including locating and isolating the target, as well as establishing legitimacy in principle. A further problem relates to the assumptions made about the identity of the target. U.S. estimates of extremely low or no civilian casualties appear to be based on a narrowed definition of "civilian," and the presumption that, unless proven otherwise, individuals killed in strikes are militants. In May 2012, *The New York Times* reported that according to unnamed Obama administration officials, the United States ". . .

in effect counts all military-age males in a strike zone as combatants . . . unless there is explicit intelligence posthumously proving them innocent."[58]

As such, there is a strong likelihood that many casualties could be civilians incorrectly categorized as combatants, especially where information on the ground may be limited. Although collateral damage is often regarded as an unfortunate side effect of war, the ethical argument of the use of drones is that it minimizes the risk of civilian casualties. As such, the truth behind the level of collateral damage is of paramount importance when endorsing the use of drones from an ethical standpoint.

Collateral Damage

Drone strikes are widely criticized for causing undue civilian casualties. A recent study estimates that of the overall number of those killed in drone strikes since 2004, 21 percent have been civilians, and only 6 percent during 2010,[59] although these numbers are highly disputed. One analysis based on open source data reported that attempts to kill 41 men resulted in the deaths of an estimated 1,147 people.[60]

Collateral damage occurs when high value individuals (HVIs) are wrongly identified or located among noncombatants. The question of how many civilians are killed in drone strikes remains highly polarized.[61] Recent UN reporting suggests that drone strikes have resulted in considerably more civilian deaths than U.S. officials have publicly acknowledged. According to UN Special Rapporteur Ben Emmerson, at least 400 civilians have been killed in Pakistan alone. In his interim report, Emmerson criticizes the United States for creating "an almost insurmountable obstacle to transparency," stating that:

> The single greatest obstacle to an evaluation of the civilian impact of drone strikes is lack of transparency, which makes it extremely difficult to assess claims of precision targeting objectively.[62]

As collateral damage minimization is a key justification for the lethal use of drones, information as to who has been killed and how many of those were intended targets is essential. Furthermore, this information needs to be publicly available if political buy-in is to be ensured not only within the United States but more widely. In addition, it should also be remembered that even where local populations escape death or injury directly through drone attacks, there are other negative security effects which need to be addressed.[63]

Other Negative Effects on Local Populations

One such example is retaliation from militant groups. In northern Pakistan, civilians have been caught in a dangerous position between local militant groups and U.S. drones. Militant groups, such as the Khorasan Mujahedin in Waziristan, pursue retaliatory attacks against local civilians they suspect of being U.S. informants. According to one report, tribal elders in North Waziristan say that most of the people killed by such militant attacks have never acted as informants, though they usually confess after beatings.[64] A further negative effect on victims is the psychological toll caused by drone attacks. Civilian deaths, injuries, displacement, and property loss caused by conflict are always traumatic for the population. Covert drone strikes take a particular toll, striking unannounced and without any public understanding of who is, and importantly, who is not a target. For victims, there is no one to recognize, apologize for, or explain their sorrow; for communities living under the constant watch of surveillance drones, there is no one to hold accountable for their fear.[65] While the United States had a practice of offering amends in the form of recognition, explanations, and monetary payments to civilians suffering losses as a result of U.S. combat operations in Vietnam, Iraq, and Afghanistan, no such amends exist for civilians harmed by U.S. drones in Pakistan, Yemen, or Somalia.[66] Furthermore, when the Center for Civilians in Conflict conducted interviews of Pakistani drone victims in 2010, all the victims believed the Pakistani or U.S. Government owed them compensation for harm resulting from drones, although not one had received any form of assistance.[67] This has and will continue to foster anti-U.S. sentiment and other second-order effects, which will be discussed later.

Perception of "Push Button Warfare" and "PlayStation Mentality"

A further ethical consideration relates to the perception of the psyche of the drone operator, who is typically based thousands of miles from the battlefield and undertakes operations entirely through computer screens and remote audio feed.[68] The operators/drone pilots face no risk of physical harm as a result of their geographical distance from the battlefield. Furthermore, media reporting has suggested that drone pilots are increasingly recruited by the military on the basis of their previous gaming skills,[69] hence the reason they reportedly are referred to colloquially as "cubicle warriors."[70] One public concern is that such cubicle warriors are likely to develop a "PlayStation" mentality to killing.[71] Such concerns are strongly denied by drone pilots. According to a former student at the U.S. Air Force 6th Reconnaissance Squadron, 2nd Lt. Zachary (last name withheld):

> Flying RPAs is nothing like playing a video game. Anyone who thinks that couldn't be more incorrect. We fly real aircraft and employ real weapons. There's nothing fake about that.[72]

Nevertheless, the terminology reportedly used deepens public concern. For example, the term "bugsplat" which is used to denote a successful attack,[73] has become official terminology used by U.S. authorities to refer to the individuals killed by a drone,[74] as the dead bodies resemble squashed bugs when rendered as pixels on a screen.[75] Another term, "squirter," refers to a person observed to run for cover in fear of a drone attack.[76] Allegedly, in the words of one drone pilot: "It's like a video game. It can get a little bloodthirsty. But it's . . . cool."[77]

Whether accurate or not, it is reporting of this kind which shapes public perception of drone strikes both within the United States and abroad. It has been argued that comments such as this emphasize the dehumanization of targets, enemy or otherwise, in drone attacks, as operators view the targets as mere objects which are expendable.[78] Critics argue that the use of such terminology, coupled with the act of killing which is remote and technology based, result in further advancing moral disengagement of those with the power to make a life and death decision.[79] However, some argue the reverse, claiming that drone operators are acutely aware of the impact of their actions as a result of technology which enables clear visual monitoring. In addition, it is argued that the high resolution images accentuate the realities on the ground, causing operators to suffer from Post-Traumatic Stress Disorder (PTSD).[80]

> . . . the carnage close-up, in real time—the blood and severed body parts, the arrival of emergency responders, the anguish of friends and family. Often he's been watching the people he kills for a long time before pulling the trigger. Drone pilots become familiar with their victims. They see them in the ordinary rhythm of their lives—with their wives and friends, with their children. War by remote control turns out to be intimate and disturbing.[81]

As of March 2015, there were approximately 1,000 drone pilots against a demand of at least 1,700. Currently, drone pilots are required to attend a year-long training program at Holloman and Randolph Air Force bases in New Mexico and Texas, respectively, before they can fly operationally. Approximately 180 pilots graduate to become drone pilots a year. However, 240 trained pilots are leaving during the same period, raising concern as to how supply can keep up with demand.[82] Although the U.S. Air Force has provided overwork and the perception of drone pilots as second rate compared to pilots of manned aircraft as reasons, PTSD is also reported to be a factor in the accelerating dropout rate of drone pilots.

It is also likely that the inappropriate language allegedly used is a coping mechanism on the part of the drone operators. Although this in itself does not justify the act of trivializing the deaths of individuals targeted by armed drones, consideration should be given that individuals who are exposed to extreme circumstances often use inappropriate humor[83] or language[84] as a defense or coping mechanism.[85] However, the U.S. military needs to raise awareness of the offense such language causes and counter the perception of trivialization, as well as provide adequate support to drone pilots who are traumatized by events on screen.

Is It Effective?

One of the arguments in favor of the use of drones is that they are effective mainly in terms of technical ability to strike intended targets. However, it is essential that the longer term impact of second and third order effects is considered fully, to ensure that they do not undermine the wider counterterrorism and COIN missions conducted by the United States. Three further key desired effects of targeted drone attacks have therefore been identified for discussion: accuracy, winning the fight, and cost.

Accuracy

For the purpose of this discussion, accuracy should be taken to mean the ability to strike the right target while minimizing collateral damage. Precision targeting requires technological ability supported by good intelligence. There are two ways in which an individual target may be identified, located, and eliminated. The first is via a method known as HVI targeting, also known as personality targeting, where an individual whose identity is known is specifically targeted. According to one study, the process involves the development of target packets based on human intelligence (HUMINT) and signals intelligence (SIGINT) which are then submitted and reviewed by a joint military board. If approved, the HVI is placed on an approved targeting list called the Joint Priority Effects List, after which concepts of operations (CONOPs) involving the HVI are developed. Once a CONOP has been approved by senior military officials, steps can be taken to remove the HVI from the battlefield.[86]

The second method is signature strikes, where unknown individuals often in groups are targeted. As the precise identity of these individuals is unknown, the individuals targeted must match a pre-identified "signature" of behavior that the United States links to militant activity or association. In other words, signature strikes differ from personality strikes in that, with the former, the specific individual targeted is known; whereas with the latter, patterns of behavior are used to determine the target. For example, if an insurgent group is suspected or known to be operating in a certain area, criteria are determined to identify and detect suspicious activities. If the activities match those criteria, the target group is eliminated.

Signature strikes are the more controversial of the two methods, but make up a significant proportion of the covert drone campaign, constituting the majority of strikes in Pakistan. Indeed, according to one unnamed U.S. official, the United States has killed twice as many "wanted terrorists" in signature strikes than in personality strikes. U.S. officials have also reported that most of the people on the CIA's kill list have been killed in signature strikes.[87] While signature strikes are clearly effective, the concern is that casting this wider net will result in other unwanted consequences, such as an increase in collateral damage.

Also, the criteria for determining suspicious behavioral patterns/signatures have come under criticism. For example, *The New York Times* quoted a senior State Department Official as saying that "three guys doing jumping jacks" would be interpreted as a possible terrorist camp.[88] Where targeting is based on biological factors, such as males between the age of 20 and 40,[89] there will always be a concern as to how many of those individuals are, in fact, civilians and whether the lack of women in the group is sufficient to assume that they are a group of insurgents. Recent experience in Afghanistan and Iraq has highlighted that insurgents and terrorists often hide among civilian populations, with many disguising themselves as women.[90] Even with on the ground in-country intelligence, identifying the right target is challenging. The challenge becomes even more so when human intelligence is lacking, and decisions need to be made based on aerial intelligence alone.

In addition, a further differentiation should be made between pre-planned attacks and dynamic targeting. Pre-planned or "deliberate" drone operations are where attacks are conducted at a scheduled time and after elaborate processes of collateral damage estimation and other steps to reduce the risk of harming civilians have taken place. In contrast, "dynamic" targeting occurs when targeting decisions are made during a short window of time, on the basis of recently received or time-sensitive information. Due to the quick turnaround time from intelligence to strike, dynamic targeting may occur without the benefits of a full collateral damage estimation and mitigation process.[91]

It should also be noted that recent reporting has shown that unsurprisingly, one negative side effect of dynamic targeting is that the risk of collateral damage increases significantly. According to one study, most collateral damage in U.S. operations occurs when collateral damage mitigation is not observed—presumably, primarily when operations are not pre-planned.[92] Even when adopting the less controversial HVI targeting methods, there are assumptions about identity which may not be correct.

The issue of target identity raises problems on several levels. Whether targeting is derived from HUMINT or SIGINT, reliability of intelligence is an issue. Faulty intelligence has sometimes led to the wrong target being struck altogether. For example, on February 21, 2010, 23 Afghan civilians were wrongly identified by a U.S. operated drone as enemy combatants and killed in airstrikes. U.S. commanders were criticized for being less than forthcoming about reporting the civilian casualties until an official investigation was launched.[93] Such incidents, again unsurprisingly, create resentment within the indigenous populations toward the United States and local governments.[94]

Furthermore, strikes can be carried out with no objective demonstration of target validity, often as a result of the challenging nature of oversight.[95] This and the adversary's demonstrated ability to replace its commanders create powerful arguments against the use of drone strikes. In fact, these types of second-order effects raise legitimate concerns about the contribution to the U.S. strategy of lethal targeting as an entire concept.

Winning the Fight

Although there is evidence to suggest that lethal targeting may be effective in terms of disrupting the enemy in the short term, the medium- and longer-term impacts of the attacks are often not fully appreciated. Short-term successes such as the killing of an enemy commander can often be short-lived as these individuals can be replaced.[96] Perhaps of more concern is that as the deaths of militants are glorified as they achieve martyr status; this, in turn, attracts new recruits to join the "cause," further exacerbating the problem. According to Mohammed al-Ahmadi, a legal coordinator for a local human rights group: "The drones are killing al-Qaeda leaders, but they are also turning them into heroes."[97]

This is particularly prevalent in the social media dominated 21st century where potential sympathizers can be mobilized from all over the globe. Targeted killings by U.S. drones play into the hands of enemy propagandists justifying their war against the United States by arguing that the use of drone strikes is an injustice from which they need to defend themselves. In 2014, the Federal Bureau of Investigation and Department of Homeland Security distributed a bulletin to law enforcement officials warning that U.S. airstrikes (to include drone attacks) could provoke retaliatory attacks on U.S. soil by IS sympathizers. IS militants have already claimed that the execution of American journalist James Foley in August 2014 was in retaliation for U.S. airstrikes in Iraq.[98]

These medium- and long-term second order effects caused by lethal targeting have often been neglected, although this is partly due to the lack of publicly available data. Too often, second order effects are cast aside as being beyond the control of friendly forces or those who conducted the strike, who are not in the location where the attacks are occurring. One reason given is that as the strikes occur in hard-to-reach places, this results in the difficulty in managing perceptions, especially over a period of time. However, this challenge does not justify taking action based on a lack of understanding.

When questioning the effectiveness of targeted killing, it is also necessary to make a distinction between counterterrorism and COIN, since they are two different policies and imply different strategies. This difference often means that what can be effective and useful as a counterterrorism tactic can be harmful from a counterinsurgency perspective. Although the two are blurred in reality, the latter traditionally involves an understanding of hearts and minds where Psychological Operations (PSYOPs) plays a key part. In other words, counterterrorism strategies, such as the elimination of a terrorist target, may prove to be counterproductive in a COIN scenario. Second order effects such as deep resentment caused among local populations for the killing of family members is a case in point.[99] With reference to U.S. drone strikes in Yemen, a lawyer in Yemen tweeted: "Dear Obama, when a US drone missile kills a child in Yemen, the father will go to war with you, guaranteed. Nothing to do with al-Qaeda."[100]

Other Yemeni observers have also argued that U.S. drone strikes create or contribute to anti-U.S. opinions and violence in general.[101] "If young men lose hope in our cause they will be looking for an alternative. And our hopeless young men are joining al-Qaeda."[102] In May 2012, a study based on interviews with government officials, tribal elders, and others in Yemen carried out by the Center for Civilians in Conflict concluded that: ". . . an unintended consequence of the attacks has been a marked radicalization of the local population."[103]

As David Kilcullen, former COIN adviser to General David Petraeus, and Andrew Exum a former U.S. Army officer in Iraq and Afghanistan then with the Center for a New American Security, noted:

> Imagine, for example, that burglars move into a neighborhood. If the police were to start blowing up people's houses from the air, would this convince homeowners to rise up against the burglars? Wouldn't it be more likely to turn the whole population against the police? And if their neighbors wanted to turn the burglars in, how would they do that, exactly? Yet this is the same basic logic underlying the drone war.[104]

As such, potential second-order effects must be considered in a complex irregular warfare environment where the differentiation between counterterrorism and COIN is often unclear. Armed drone strikes cannot be conducted or evaluated in isolation, which adds to the challenge of determining their overall effectiveness.

Cost

Cost in the context of the use of armed drones can be measured in terms of both fiscal and political cost. In terms of the first, drones are an inexpensive option compared to other types of aircraft used in air-strikes which require a pilot to be on board. In terms of the aircraft itself, as configuration of the drone is not dictated by the human, it can be of any size. The lack of a pilot and passengers also negates the need for support systems such as pressurized cabins, further reducing cost.

Drones also satisfy the political dimension by reducing or eliminating the need for U.S casualties. One desired result is to achieve the mission with minimal casualties, referring both to military personnel and to innocent civilians caught in the crossfire. Here, one advantage of "remote" warfare is that the lack of soldiers on the ground results in the prevention of military casualties, which will be politically popular within the United States in that it can be seen to be saving U.S. lives. This may be particularly relevant when engaged in irregular warfare, a concept that is complex and one which the general public does not always fully understand or support.

An added practical operational benefit, which has both financial and political implications, is access. Ground operations, even small tactical strikes using Special Forces, are not only costly in monetary terms but extremely dangerous to troops. As drones are able to penetrate the most remote locations through their ability to fly for long hours without the need to refuel, as well as to access air space and terrain that would not be considered safe for a piloted aircraft to approach, this can be attractive from both a financial and political viewpoint. At first glance, the benefits are self-evident. However, remote warfare incurs other costs which also need to be considered. For example, the "cowardly" nature of drone warfare resulting from the lack of risk of physical harm to the drone pilots works against U.S. efforts to maintain a positive image.[105] Adversary groups such as ISIS have commented: "Don't be cowards and attack us with drones. Instead, send your soldiers, the ones we humiliated in Iraq."[106]

Importantly, this perception is not limited to adversary organizations. David Kilcullen has also expressed his concern that ". . . using robots from the air . . . looks both cowardly and weak."[107] Kilcullen's comment is based purely on practical grounds, as he warned of the use of drone strikes backfiring with the potential to create more enemies than they eliminate. For example, in the tribal areas of Pakistan, where U.S. drone strikes have significantly weakened al-Qaeda's capabilities, an unintended consequence of the attacks reported has been a marked radicalization of the local population.[108] The evidence of radicalization emerged in more than 20 interviews with tribal leaders, victims' relatives, human rights activists, and officials from four provinces in southern Yemen where U.S. strikes have targeted suspected militants. They described a strong shift in sentiment toward militants affiliated with the transnational network's most active wing, alQaeda in the Arabian Peninsula (AQAP).[109]

Other critics, notably remote from the battlefield, describe the use of technology driven warfare over conventional face-to-face methods as not only "cowardly" but also unfair and dishonorable.[110] Peter Singer of the New America Foundation has questioned whether drone warfare can even be described as war and raises concerns regarding accountability. Singer argues that a president who sends someone's son or daughter into battle has to justify it publicly, as does the congress responsible for appropriations and a declaration of war. But, if no one has children in danger, Singer questions whether drone warfare can be considered to be warfare at all.[111]

This perception of drone pilots as "cowards" and second rate compared to their manned pilot counterparts is reported to be another factor affecting their retention.[112] Taking all these factors into account, the political benefits to the United States, namely that drone warfare negates the need to justify human losses to voters, are not straightforward, and require more careful consideration.

FUTURE USE OF DRONES FOR TARGETED KILLINGS

Despite the controversies surrounding drone warfare, in particular targeted killings by armed drones, demand for drone operations continues to increase in the United States. According to Air Force statistics, Predators and Reapers flew 369,913 flight hours in 2014, a figure six times higher than in 2006.[113] Furthermore, the Pentagon is asking Congress for $904 million in 2016 to buy 29 Reapers, more than double the number it sought in 2015.[114]

Speaking at a defense conference in Washington in March 2015, Deputy Secretary of Defense Robert O. Work confirmed that "Commanders' appetite for drones remains very, very high and continues to outstrip our supply."[115] Furthermore, in a prepared statement in March 2015 to the House Armed Services Committee, General Lloyd Austin, the overall commander of U.S. forces in the Middle East and Afghanistan, also highlighted the reliance of the U.S. military on drones and the need to address shortages in supply, stating that drones, and in particular the video footage they provide ". . . had become fundamental to almost all battlefield maneuvers."[116] In addition to the continued demand for drones, technological advancements will inevitably lead to further development and increased proliferation of drone technology, in the absence of direct intervention to prevent such development. There is currently no indication that such intervention can be expected in the near future.

However, there are two areas of concerns in relation to future drones. The first relates to the development of Nano drones and the second, autonomous drones. Nano drones the size of insects could be used for targeted killings using poison or other methods which do not require a large payload.[117] A prototype "hummingbird drone," capable of flying at 11 miles per hour and perching on a windowsill, was unveiled in February 2011.[118] In the same year, media reporting warned of fully autonomous drones, capable of identifying and eliminating a target without direct human intervention, which were being prepared for deployments by the United States.[119] Some have expressed concern that such autonomous drones potentially represent ". . . the greatest challenge for *jus in bello* since the development of chemical warfare."[120]

Proposals for autonomous UAVs in general are fraught with ethical dilemmas. At present, the technology is insufficiently developed to be operationalized. However, it is likely that sometime in the future autonomous UAVs will become a reality as advancements in facial recognition technologies reach the stage where target recognition without human intervention becomes possible. It is essential that the ethical implications of the use of such technologies are debated in full and resolved, including issues surrounding accountability. For example, if an autonomous drone were to target a school bus instead of a tank, systems must be in place to ensure that responsibility will not simply be devolved and blame placed exclusively on systems failures and technology.

CONCLUSION

Drones make warfare cheaper and easier, as well as more efficient, by transcending human limitations. Furthermore, a drone is dispensable and incurs much less political cost when shot down or "killed" than a conventional aircraft with pilot. But the use of drones for targeted killings has generated significant controversy. While supporters claim that drone warfare is not only legal but ethical and wise, others have suggested that drones are prohibited weapons under IHL because they cause, or have the effect of causing indiscriminate killings of civilians, such as those in the vicinity of a targeted person.[121] Questions have also been raised over the methods used for locating and eliminating targets. The use of signature strikes in particular has been criticized for being unable to distinguish sufficiently between a legitimate target and an innocent civilian, as methods used only take into consideration basic biological factors such as age and gender in identifying a potential militant.

Leaving aside technical aspects of accurate targeting, there is doubt as to the medium and longer term overall effectiveness of conducting drone missions, because they cause resentment among local populations and give fuel to the enemy cause. In particular, they provide the enemy with freshly-converted supporters who previously might have been undecided, neutral or even positively disposed toward the United States.

This anti-American sentiment is not necessarily confined to the victims of drone attacks in countries affected by the strikes. Following the attacks of 9/11, cooperation between the United States and its allies on counterterrorism has increased substantially. This cooperation appeared based on shared objectives and values. For example, in 2004, the European Union (EU) and the United States adopted a Declaration on Combating Terrorism that spelled out the objectives of their counterterrorism cooperation. The declaration stated that U.S.-EU counterterrorism cooperation would be in keeping with human rights and the rule of law. Since then, however, the United States has expanded its counterterrorism tactics beyond what many in the EU would consider the limits of international law.[122]

The use of armed drones is now added to a list of controversial U.S. counterterrorism tactics such as the maintenance of the detention center in Guantanamo Bay, Cuba, where suspected terrorists have been held on an often dubious legal basis; the use of interrogation systems bordering torture, such as waterboarding; and extraordinary renditions, whereby terrorist suspects abducted in third countries were then transferred to states where a lax legal system and oversight providing no guarantee against torture or inhuman treatment could be exploited.[123] The benefit of these tactics must be weighed against the reputation of the United States among its allies, as well as the key question of the extent to which they are having a positive impact in countering terrorism, or whether they are, in fact, doing more to fuel terrorism.

There are no easy answers. Given the enormously complex and multidimensional nature of terrorism, any action to counter it must be considered with extreme caution, since action taken to address one aspect of the problem in isolation may reveal or create additional problems. As for drones, while further advancement in relevant technologies is inevitable, it must also be recognized that technology in general is advancing at a pace that outstrips the political, legal and ethical frameworks upon which coexistence and cooperation with global partners is built.

In tackling the threat of global terrorism, it is often too easy for practitioners to become focused on their specific problem set to the point where the wider consequences of their action are forgotten or put aside. This is dangerous, as such second order effects may undo well-intentioned actions and exacerbate the original problem. Targeting methods such as signature strikes are a case in point, where casting a wider net to eliminate a terrorist or a group of terrorists may result in significant collateral damage and be detrimental to the wider mission. It must be remembered that methods used by terrorists include sacrificing "parts" for the benefit of the "whole." The U.S. administration must be careful to ensure that it cannot be accused of doing the same.

RECOMMENDATIONS

Recommendations include the following:

1) **Investigate.** Conduct a review of civilian casualties both in and out of declared theaters of armed conflict. The review should contain information on the number of civilian deaths and injuries, specifying

whether the victim was male or female, adult or child. In addition, the larger impact on civilian communities, including destruction of homes and displacement, and retaliatory violence by local groups must be included as part of the analysis.

2) **Transparency**. The outcome of the fact-finding investigations should be made public, except where operational considerations preclude this. In such situations, the government should at a minimum explain its decision.

3) **Consequence Management**. Best practices and lessons from Afghanistan regarding civilian casualty consequence management should be applied to other U.S. Government efforts, including operations outside declared theaters of armed conflict.

4) **Targeting (Intelligence)**. Conduct a review to determine the adequacy of standards for the identification of targets, including the reliability of "signatures," and the sufficiency of intelligence sources and analysis especially where there is limited U.S. ground presence.

5) **Targeting (Classification)**. Review the process for classifying casualties as enemy combatants versus civilians in operations outside declared theaters of armed conflict.

6) **Pilots**. Continue to verify the validity and relevance of drone pilot education in the legal and ethical dimension of armed attacks. Ensure continued provision of psychological support to personnel where necessary.

7) **Perception Management**. In addition to creating transparency, ensure that the message (namely why, what, and how) is effectively communicated both domestically and abroad, to illustrate that U.S. armed drone strikes are indeed legal, ethical, and accurate.

End Notes

[1] The Security Impact of Drones: Challenges and Opportunities for the UK, Birmingham Policy Commission Report, Birmingham, UK, October 2014.

[2] "US Unmanned Systems Integrated Roadmap (Fiscal Years 2009–2034)," Washington, DC: U.S. Department of Defense, 2009, p. 2, available from www.acq.osd.mil/sts/docs/ UMSIntegrated Roadmap2009.pdf, accessed December 17, 2014.

[3] Alain De Neve, "Armed drones: What technical revolution? What political and ethical implications?" Brussels, Belgium: Centre for Security and Defence Studies of the Royal Higher Institute for Defence, October 22, 2013, available from www.irsd.be/website/ images/livres/enotes/en/eNote11EN.pdf, accessed December 18, 2014.

[4] Peter Bergen and Katherine Tiedemann, "Hidden war, there were more drone strikes—and far fewer civilians killed," Washington, DC: New America Foundation, December 22, 2010, available from newamerica.net/node/41927, accessed January 2, 2015.

[5] W. J. Hennigan, "New drone has no pilot anywhere, so who's accountable?" Los Angeles Times, January 26, 2012, available from www.latimes.com/business/la-fi-auto-drone-20120126,0,740306. story, accessed December 20, 2014.

[6] Afsheen John Radsan, "Loftier standards for the CIA's remote-control killing," Statement for the House Subcommittee on National Security & Foreign Affairs, in Legal Studies Research Paper Series, Accepted Paper No. 2010–11, St Paul, MN: William Mitchell College of Law, May 2010, available from papers.ssrn.com/sol3/papers.cfm?abstract_id=1604745, accessed December 20, 2014.

[7] Birmingham Policy Commission Report, p. 15.

[8] Based on the design for the ubiquitous DH82 Tiger Moth trainer, the Queen Bee first flew in 1935. It was capable of flying at 17,000 feet, and had a range of 300 miles at over 100 mph. A total of 380 Queen Bees served as target drones in the Royal Air Force and the Royal Navy until they were retired in 1947.

[9] House of Commons Defence Committee, "Remote Control: Remotely Piloted Air Systems—Current and Future UK use—Tenth Report of Session 2013-14," Vol. 1, London, UK: The Stationery Office Limited, 2014, p. 11.

[10] Birmingham Policy Commission Report.

[11] "Drone Journalism and the Law," Chapel Hill, NC: UNC Centre for Media Law and Policy, available from https://medialaw. unc.edu/resources/drone-journalism/, accessed May 14, 2015. See also Jennifer O'Mahony, "The brave new world of 'drone journalism'," The Telegraph, June 13, 2013, available from www.telegraph.co.uk/technology/news/10129485/The-brave-new-world-of-drone-journalism.html, accessed May 14. 2015.

[12] "Northern Ireland police to use drones at G8 summit: Police force want to buy two remote-controlled aircraft to assist with security at this summer's economic summit," The Guardian, March 15, 2013, available from www.theguardian.com/uk/2013/mar/15/northern-ireland-drones-g8-summit, accessed January 10, 2015.

[13] "German railways to test anti-graffiti drones," BBC News, May 27, 2013, available from www.bbc.co.uk/news/worldeurope-22678580, accessed January 10, 2015.

[14] "Yo! Sushi: The world's first flying waiter—the 'iTray'," June 12, 2013, available from https://www.youtube.com/watch?v=y9RKXO1rr7g and metro.co.uk/2013/06/11/video-worldsfirst-flying-tray-at-sushi-restaurant-yo-sushi-3836390/, accessed January 10, 2015.

[15] "Take Off For Sushi Restaurant's Flying Tray," Sky News, June 10, 2013, available from news.sky.com/story/1101379/take-offfor-sushi-restaurants-flying-tray, accessed January 10, 2015.

[16] "Saving the rhino with surveillance drones," The Guardian, December 25, 2012, available from www.theguardian.com/environment/2012/dec/25/saving-the-rhino-with-surveillance-drones, accessed January 10, 2015.

[17] Winnie Agbonlahor, "The Secret Life of Drones," Civil Service World, July 5, 2013.

[18] Harold Hongju Koh, legal adviser of U.S. Department of State, The Obama Administration and International Law, speech at the annual meeting of the American Society of International Law, Washington, DC, March 25, 2010, available from www.state.gov/s/l/releases/remarks/139119.htm, accessed January 11, 2015.

[19] United Nations General Assembly, Sixty-eighth Session, Report of the Special Rapporteur on Extrajudicial, Summary, or Arbitrary Executions, "Protecting People: the right to life," A/68/382, September 13, 2014, Sec. Para. 30.

20 Rory Carroll, "The Philosopher Making the Moral Case for US Drones," The Guardian, available from www.theguardian.com/world/2012/aug/02/philosopher-moral-case-drones, accessed December 20, 2014.

21 New York City Bar, "The Legality under International Law of Targeted Killings by Drones Launched by the United States," Committee on International Law, June 2014.

22 Geneva Convention III, Art. 2.

23 Paul Rogers, "America's military: failures of success," Open Democracy, May 12, 2011.

24 Eric Schmitt and Thom Shanker, Counterstrike: The Untold Story of America's Campaign Against al-Qaeda, New York: Times Books, 2011, p. 245.

25 International Committee of the Red Cross Website, available from https://www.icrc.org/en/war-and-law/ihl-other-legal-regmies/jus-in-bello-jus-ad-bellum, accessed December 20, 2014.

26 International Committee of the Red Cross, "International Humanitarian Law: Answers to your Questions," October 2002, available from www.redcross.org/images/MEDIA_Custom ProductCatalog/m22303661_IHL-FAQ.pdf, accessed May 14, 2015.

27 Jasmine Moussa, "Can 'jus ad bellum' override 'jus in bello'? Reaffirming the separation of the two bodies of law," International Review of the Red Cross, Article No. 852, December 31, 2008.

28 Francois Bugnion, "Guerre juste, guerre d'aggression et droit international humanitaire" ("Just war, war of aggression, and international humanitarian law"), International Review of the Red Cross, Vol. 84, 2002, p. 528.

29 Moussa.

30 Ibid.

31 S. Casey-Maslen "Pandora's box? Drone strikes under jus ad bellum, jus in bello, and international human rights law," International Review of the Red Cross, Vol. 94, No. 886, Summer 2012, p. 597.

32 Jane Mayer, "The Predator War: What are the risks of the C.I.A.'s covert drone program?" The New Yorker, October 26, 2009, available from www.newyorker.com/magazine/2009/10/26/the- predator-war, accessed May 5, 2015.

33 Avery Plaw, Targeting Terrorists: A License to Kill? Aldershot, UK, and Burlington, US: Ashgate Publishing, 2008; Jens David Ohlin, Is Jus in Bello in Crisis? No. 3, Ithaca, NY: Cornell Law Faculty Publications, 2013, available from scholarship.law.cornell.edu/cgi/viewcontent.cgi?article=2475&context=facpub, accessed April 23, 2015.

34 If the attack that killed al-Libi was part of an armed conflict, the lawfulness of the attack depends on its compliance with the rules of IHL. What al-Libi was doing at the time he was killed (and who was with him) would be relevant in assessing whether the attack was proportionate, and whether the second strike complied with the prohibition of targeting civilians not directly participating in hostilities who are involved in rescuing the wounded and the prohibition of targeting the wounded themselves.

35 Non-International Armed Conflict. In situations of NIAC, the relevant bodies of law that come into play are LoAC/IHL, international human rights law (IHRL) and each State's domestic law. See "Procedural Safeguards for Security Detention in Non-International Armed Conflict," London, UK: Chatham House & ICRC, September 22-23, 2008, available from https://www.chathamhouse.org/sites/files/chathamhouse/field/field_document/Meeting%20 Summary%20Procedural%20Safeguards%20for%20Security%20Detention%20in%20 Non-International%20Armed%20Conflict.pdf.

[36] Benjamin Wittes, "Amnesty International Responds," Law-fare, October 25, 2013, available from www.lawfareblog.com/2013/10/amnesty-international-responds/, accessed January 20, 2015.

[37] See section, in this monograph, that follows on efficacy entitled, "Is it Effective?" for methods used for targeting.

[38] Assassinations generally refer to the elimination of political officials, and were banned in the United States in 1981. In 1975, the U.S. Senate Select Committee chaired by Senator Frank Church (the Church Committee) reported on several CIA assassinations, including efforts to assassinate Fidel Castro. Criticism in this report resulted in a proposal banning assassinations. This was later incorporated into Executive Order 12333 signed by President Ronald Reagan in 1981, which is still in effect today. Executive Order 12333 is available at https://www.cia.gov/about-cia/eo12333.html.

[39] Whitley R. P. Kaufman, "Rethinking the ban on assassination: just war principles in the age of terror," In Michael Brough, John Lango and Harry van der Linden, eds., Rethinking the Just War Tradition, New York: State university of New York, 2007, pp. 171-182.

[40] Yael Stein, "By any name illegal and immoral," Ethics & International Affairs, Vol. 17, No. 1, 2003, pp. 127-136.

[41] Greg Miller and Bob Woodward, "Secret memos reveal explicit nature of U.S., Pakistan agreement on drones," The Washington Post, October 24, 2013, available from www.washingtonpost. com/world/national-security/top-pakistani-leaders-secretly-backed-cia-drone-campaign-secret-documents-show/2013/10/23/15e6b0d8-3beb-11e3-b6a9-da62c264f40e_story.html, accessed May 5, 2015.

[42] Asif Shahzad and Steven R. Hurst, "Pakistan's Drone Strike Victims Accuse U.S. Of Double Standard," Huffington Post, May 1, 2015, available from www.huffingtonpost.com/2015/05/01/pakistan-drone-strike-double-standard_n_7188644.html?utm_hp_ref=politics&ir=Politics, accessed May 3, 2015. See also William Helbling, "Pakistan top court denies petition challenging US drone strikes," The Jurist, April 13, 2015, available from jurist.org/paperchase/2015/04/pakistan-top-court-denies-petition-challenging-usdrone-strikes.php, accessed May 3, 2015.

[43] John Brennan, Assistant to the President for Homeland Security and Counterterrorism, "The Ethics and Efficacy of the President's Counterterrorism Strategy," speech, Wilson Center, Washington, DC, April 30, 2012, available from www.wilsoncenter. org/event/the-efficacy-and-ethics-us-counterterrorism-strategy.

[44] New York City Bar, "The Legality under International Law of Targeted Killings by Drones launched by the United States," New York City, June 2014, p. 3.

[45] See Legality of the Threat or Use of Nuclear Weapons, Advisory Opinion, 1996 I.C.J. 226, 41, July 8.

[46] The Geneva Convention 1949, Additional Protocol I, Art. 51, 5(b).

[47] Helen Duffy, The "War on Terror" and the Framework of International Law, Cambridge, UK: Cambridge University Press, 2005, pp. 18-19, 231-235.

[48] Noam Neuman, "Applying the Rule of Proportionality: Force Protection and Cumulative Assessment in International Law and Morality," 7 Y.B. International Humanitarian Law, 2004, pp. 79, 81, 98-99.

[49] In Prosecutor v. Galić, the Trial Chamber observed that proportionality should be considered under the lens of "whether a reasonably well-informed person in the circumstances of the actual perpetrator, making reasonable use of the information available to him or her, could have expected excessive civilian casualties to result from the attack," Prosecutor v. Galić, Case No. IT-98-29-T, Judgement and Opinion, 58, ICTY, December 5, 2003.

50 "The Legality under International Law of Targeted Killings by Drones Launched by the United States," p. 170.

51 Ibid.

52 Testimony of Brian Michael Jenkins of RAND Corporation, Al Qaeda After Bin Laden: Hearing Before the House Armed Services Committee, Subcommittee on Emerging Threats and Capabilities, 112th Cong., 2011.

53 Casey-Maslen, p. 597.

54 Bradley Strawser is the author of Killing by Remote Control: The Ethics of an Unmanned Military.

55 Bradley J. Strawser in Rory Carroll, "The philosopher making the moral case for US drones," The Guardian, August 2, 2012, available from www.theguardian.com/world/2012/aug/02/philosopher-moral-case-drones, accessed April 23, 2015.

56 Ibid.

57 See section, in this monograph, that follows, "Is it Effective?" with respect to the accuracy of drone strikes.

58 Jo Becker and Scott Shane, "Secret 'Kill List' Proves a Test of Obama's Principles and Will," The New York Times, May 29, 2012.

59 Peter Bergen and Katherine Tiedemann, The Year of the Drone, Washington, DC: New America Foundation, available from foreignpolicy.com/2010/04/23/the-year-of-the-drone/, accessed December 14, 2014.

60 Spencer Ackerman, "41 men targeted but 1,147 people killed: US drone strikes—the facts on the ground," The Guardian, November 24, 2014, available from www.theguardian.com/us-news/2014/nov/24/-sp-us-drone-strikes-kill-1147, accessed April 23, 2015.

61 Chris Woods, "Analysis: CNN expert's civilian drone death numbers don't add up," London, UK: Bureau of Investigative Journalism, July 17, 2012, available from www.thebureauinvestigates.com/2012/07/17/analysis-cnn-experts-civilian-drone-death-numbers-dont-add-up/, accessed December 15, 2014.

62 UN General Assembly, Report of the Special Rapporteur on the promotion and protection of human rights and fundamental freedoms while countering terrorism, September 18, 2013. p. 11, available from msnbcmedia.msn.com/i/msnbc/sections/news/UN_ Drones_Report.pdf, accessed January 12, 2015.

63 "The Civilian Impact of Drones: Unexamined Costs, Unanswered Questions," New York: Columbia Law School, Center for Civilians in Conflict, p. 21.

64 Alex Rodriguez, "Pakistani Death Squads Go after Informants to US Drone Program," The Los Angeles Times, December 28, 2011; Jane Mayer, "The Predator War: What are the risks of the C.I.A.'s covert drone program?" The New Yorker, October 26, 2009.

65 "The Civilian Impact of Drones," p. 24.

66 Ibid., p. 26.

67 Ibid.

68 "UN official criticises US over drone attacks," BBC News, June 2, 2010, available from www.bbc.co.uk/news/10219962, accessed May 14, 2015.

69 "War Games: Documentary shows how U.S. recruited gamers to fly drones into Pakistan," New York Times Tribune, February 2, 2015, available from tribune.com.pk/story/831710/war-games-documentary-shows-how-us-recruited-gamers-to-fly-drones-in-to-pakistan/, accessed May 3, 2015.

70 Tonje Schei, Lars Løge, Jonathan Lie, Charlie Phillips, Noah Payne-Frank, and Jacqui Timberlake, "Drone wars: the gamers recruited to kill"—video, The Guardian Documentary, available from www.theguardian.com/news/video/2015/feb/02/drone-wars-

gamers-recruited-kill-pakistan-video, accessed May 3, 2015; Dan Pearson, "War Games: the link between gaming and military recruitment," Gamesindustry.biz, February 2, 2015, available from www.game sindustry.biz/articles/2015-02-02-the-military-recruitment-of-gamers, accessed May 3, 2015.

[71] "UN official criticises US over drone attacks."

[72] Second-Lieutenant Logan Clark, 49th Wing Public Affairs, "18X pilots learn RPAs first," U.S. Air Force, Holloman Air Force Base Website, September 2, 2012, available from www.holloman. af.mil/news/story.asp?id=123289389, accessed May 17, 2015.

[73] Sophia Saifi, "Not a 'bug splat:' Artists give drone victims a face in Pakistan," CNN News, April 9, 2014, available from edition.cnn.com/2014/04/09/world/asia/pakistan-drones-not-a-bug-splat/, accessed May 14, 2015. See also Leo Benedictus, "The artists who are giving a human face to the US's 'bug splat' drone strikes," The Guardian, April 7, 2014, available from www.theguardian.com/world/shortcuts/2014/apr/07/artists-give-human-face-drones-bugsplat-pakistan, accessed May 14, 2015.

[74] Rob Williams, "Giant 'Not A Bug Splat' art installation takes aim at Pakistan's predator drone operators," The Independent, April 8, 2014, available from www.independent.co.uk/news/world/asia/giant-not-a-bug-splat-art-installation-takes-aim-at-pakistans-predator-drone-operators-9246768.html, accessed May 30, 2015.

[75] E. Robinson, "Assassination by robot: Are we justified?" The Washington Post, 2011, available from www.washingtonpost. com/opinions/assassination-byrobot-are-we-justified/2011/06/30/AGp0DlsH_story.html, accessed November 3, 2014.

[76] J. Mayer, "The Predator War—What are the risks of the CIA's covert drone program?" The New Yorker, 2009.

[77] P. W. Singer, Wired for War: The Robotics Revolution and Conflict in the 21st Century, London, UK: Penguin, 2010a, pp. 308-309.

[78] Sikander Ahmed Shah, International Law and Drone Strikes in Pakistan: The Legal and Socio-political Aspects, London, UK: Routledge, November 2014, p. 209.

[79] Ann Rogers, "Investigating the Relationship Between Drone Warfare and Civilian Casualties in Gaza," Journal of Strategic Security, Vol. 7, No. 4, 2014, pp. 94-107, available from scholarcommons.usf.edu/jss/vol7/iss4/8, accessed May 14, 2015. See also Tyler Wall and Torin Monahan, "Surveillance and violence from afar: The politics of drones and liminal security-scapes," Theoretical Criminology, August 15, 2011, pp. 239-254, available from tcr. sagepub.com/content/15/3/239.abstract, accessed May 14, 2015.

[80] Statement of Rosa Brooks, "The Constitutional and Counterterrorism Implications of Targeted Killings," Hearing Before the Senate Judiciary Subcommittee on the Constitution, Civil Rights, and Human Rights, 113th Cong., 2013.

[81] Mark Bowden, "The Killing Machines," The Atlantic, August 14, 2013, available from www.theatlantic.com/magazine/print/2013/09/the-killing-machines-how-to-think-about-drones/309434/, accessed January 23, 2015.

[82] Pratap Chatterjee, "American Drone Operators Are Quitting in Record Numbers," The Nation, March 5, 2015, available from www.thenation.com/article/200337/american-drone-operators-are-quitting-record-numbers, accessed May 17, 2015.

[83] Jon Kelly, "Why do people tell sick jokes about tragedies?" BBC News, March 18, 2011, available from www.bbc.co.uk/news/magazine-12775389, accessed May 29, 2015.

[84] P. R. McCrae, "Situational determinants of coping responses: Loss, threat and challenge," Journal of Personality and Social Psychology, Vol. 46, 1984, pp. 919-928.

[85] Raija-Leena Punamäki, Katri Kanninen, Samir Qouta, and Eyad, "The role of psychological defences in moderating between trauma and post-traumatic symptoms among Palestinian men," International Journal of Psychology, Vol. 37, Issue 5, 2002, pp. 289-296.

[86] Luke A. Olnet, "Lethal Targeting Abroad: Exploring longterm effectiveness of armed drone strikes in overseas contingency operations," Georgetown University MA Thesis, Washington, DC, March 11, 2015.

[87] Greg Miller, "C.I.A. Seeks New Authority to Expand Yemen Drone Campaign," The Washington Post, April 18, 2012.

[88] Adrianna Huffington, "'Signature strikes' and the President's Empty Rhetoric on Drones," Huffington Post, July 10, 2013, available from www.huffingtonpost.com/arianna-huffington/signature-strikes-and-the_b_3575351.html, accessed May 30, 2015. See also Jo Becker and Scott Shane, "Secret 'Kill List' Proves a Test of Obama's Principles and Will," The New York Times, May 29, 2012, available from www.nytimes.com/2012/05/29/world/obamasleadership-in-war-on-al-qaeda.html?pagewanted=all, accessed May 31, 2015.

[89] Tara McKelvey, "A Former Ambassador to Pakistan Speaks Out," The Daily Beast, November 20, 2012, available from www.thedailybeast.com/articles/2012/11/20/a-former-ambassador-to-pakistan-speaks-out.html, accessed May 31, 2015.

[90] "Afghan-Led Force Captures Taliban Commander, Suspected Insurgents Dressed as Women," ISAF Joint Command— Afghanistan, 2010-07-CA-195, available from www.rs.nato.int/article/isaf-releases/afghan-led-force-captures-taliban-commander- suspected-insurgents-dressed-as-women.html, accessed June 1, 2015.

[91] "The Civilian Impact of Drones," p. 11.

[92] Gregory S. McNeal, "US Practice of Collateral Damage Estimation and Mitigation," Social Science Research Network, November 9, 2011, available from papers.ssrn.com/sol3/papers.cfm?abstract_id=1819583, accessed February 20, 2015.

[93] Robert H. Reid, "Officers reprimanded in fatal drone strike," Army Times, May 29, 2010, online edition, available from www.armytimes.com/news/2010/05/ap_afghanistan_airforce_ drone_052910/, accessed November 7, 2014.

[94] George C. Wilson, "The True Cost of Assassinations," Congress Daily, September 13, 2010.

[95] Micah Zenko, "Reforming U.S. Drone Strike Policies," Council Special Report No. 65, New York: Council on Foreign Relations, Center for Preventive Action, January 2013.

[96] "The Civilian Impact of Drones," p. 23.

[97] Theodore Karasik, "The Drone Doctrine in Yemen—Understanding the Whole Picture," Arabian Aerospace, July 24, 2012.

[98] Catherine Herridge, "FBI, DHS bulletin warns of retaliation for airstrikes against ISIS," Fox News, August 26, 2014, available from www.foxnews.com/politics/2014/08/26/fbi-dhs-bulletinwarns-retaliation-for-airstrikes-against-isis/, accessed April 23, 2015.

[99] "The Civilian Impact of Drones," p. 23.

[100] Haykal Bafana, Twitter post, May 11, 2012, available from https://twitter.com/BaFana3/statuses/200930818816880640; Mothana, "How Drones Help Al Qaeda," The New York Times, June 13, 2012.

[101] "The Civilian Impact of Drones," p. 23.

[102] Ghaith Abdul-Ahad, "Yemenis Choose Jihad over Iranian Support," The Guardian, May 10, 2012; Micah Zenko, "Escalating America's Third War in Yemen," Washington, DC: Council on Foreign Relations, May 14, 2012.

[103] Sudarsan Raghavan, "In Yemen, US Airstrikes Breed Anger, and Sympathy for Al Qaeda," The Washington Post, May 29, 2012. See also commentary in "The Civilian Impact of Drones," p. 23.

[104] David Kilcullen and Andrew McDonald Exum, "Death from Above, Outrage Down Below," The New York Times, May 16, 2009; Noah Shachtman, "Call Off Drone War, Influential US Adviser Says," Wired, February 10, 2009 Civilian Conflict Report, p. 23.

[105] George Monbiot, "With its deadly drones, the US is fighting a coward's war," The Guardian, January 30, 2012, available from www.theguardian.com/commentisfree/2012/jan/30/deadlydrones-us-cowards-war, accessed May 14, 2015.

[106] Andres Johnson, "ISIS to U.S.: "Don't be Cowards and Attack us With Drones. Send your Soldiers," National Review, August 8, 2014, available from www.nationalreview.com/corner/384981/isis-us-dont-be-cowards-and-attack-us-drones-send-your-soldiers-andrew-johnson, accessed May 17, 2015.

[107] Doyle McManus, "U.S. drone attacks in Pakistan 'backfiring,' Congress told," The Los Angeles Times, May 3, 2009, available from articles.latimes.com/2009/may/03/opinion/oe-mcmanus3, accessed May 17, 2015.

[108] Sudarsan Raghavan, "In Yemen, U.S. airstrikes breed anger, and sympathy for al-Qaeda," The Washington Post, May 29, 2012, available from www.washingtonpost.com/world/middle_east/in-yemen-us-airstrikes-breed-anger-and-sympathy-for-al-qaeda/2012/05/29/gJQAUmKI0U_story.html, accessed May 17, 2015.

[109] Ibid.

[110] Steven John Thompson, "Global Issues and Ethical Considerations in Human Enhancement Technologies," IGI Global, 2014, Hershey, PA, p. 147.

[111] "Robots at War: Drones and Democracy," The Economist, October 1, 2015, available from www.economist.com/blogs/babbage/2010/10/robots_war, accessed May 17, 2015.

[112] Pratap Chatterjee, "American Drone Operators Are Quitting in Record Numbers," The Nation, March 5, 2015, available from www.thenation.com/article/200337/american-drone-operators-are-quitting-record-numbers, accessed May 17, 2015.

[113] Craig Whitlock, "How crashing drones are exposing secrets about U.S. Operations," Washington Post, March 25, 2015, available from www.washingtonpost.com/world/national-security/how-crashing-drones-are-exposing-secrets-about-uswar-operations/2015/03/24/e89ed940-d197-11e4-8fce-3941fc548f1c_story.html?utm_source=Sailthru&utm_medium=email&utm_ term=%2 ASituation%20Report&utm_campaign=Sit%20Rep%20March%2026%202015, accessed April 8, 2015.

[114] Ibid.

[115] Ibid.

[116] Ibid.

[117] Nils Melzer, "Targeted Killings in International Law," Oxford Monographs in International Law, Oxford, UK: Oxford University Press, 2008, pp. 3–4.

[118] Elisabeth Bumiller and Thom Shanker, "War evolves with drones, some tiny as bugs," The New York Times, June 19, 2011, available from www.nytimes.com/2011/06/20/world/20drones. html?pagewanted=1&_r=1&ref=unmannedaerialvehicles.

[119] W. J. Hennigan, "New drone has no pilot anywhere, so who's accountable?" The Los Angeles Times, January 26, 2012, available from www.latimes.com/business/la-fi-autodrone-20120126,0,740306.story, accessed January 14, 2015.

[120] Emma Slater, "UK to spend half a billion on lethal drones by 2015," The Bureau of Investigative Journalism, November 21, 2011, available from https://www.thebureauinvestigates.com/2011/11/21/britains-growing-fleet-of-deadly-drones/, accessed December 29, 2014.

[121] Philip Alston, Study on Targeted Killings (A/HRC/14/24/Add.6), Report of the Special Rapporteur on Extrajudicial, Summary, or Arbitrary Executions to the UN Human Rights

Council, May 28, 2010, p. 24, available from www2.ohchr.org/english/bodies/hrcouncil/docs/14session/A.HRC.14.24.Add6.pdf, accessed April 23, 2015.

[122] Nathalie Van Raemdonck, "Vested Interest or Moral Indecisiveness? Explaining the EU's Silence on the US Targeted Killing Policy in Pakistan," Istituto Affari Internazionali Working Papers 12-05, Rome, Italy: Istituto Affari Internazionali, March 2012, p. 2.

[123] Ibid.

ABOUT THE AUTHOR

SHIMA D. KEENE is a British academic and practitioner specializing in matters relating to national and international security. She is a Director of the Conflict Studies Research Centre, Oxford, as well as Director of the Security Economics Programme at the Institute for Statecraft, London, United Kingdom (UK). Dr. Shima is also a Deployable Civilian Expert (DCE)[1] and a member of the UK Government's Civilian Stabilisation Group (CSG)[2] specializing in Intelligence and Security Sector Reform which sits under the Security and Justice Function of the CSG. She is a former Senior Research Fellow and Advisor at the Advanced Research and Assessment Group, Defence Academy of the UK, and Special Advisor to the UK Ministry of Defence (MoD), where she had responsibility for assessment and recommendations for the development of financial counterinsurgency strategies in Afghanistan. Dr. Keene advises and works closely with several British and international organizations on a number of topics relating to national and international security including the MoD, various UK Government departments and law enforcement agencies, the Organization for Security and Co-operation in Europe, the Council of Europe, the Global Futures Forum, the North Atlantic Treaty Organization (NATO), U.S. Government departments and law enforcement agencies, and various global private sector organizations. Dr. Shima has 25 years of practitioner experience in a number of industries to include finance, defense, security, and telecommunications in both the public and private sectors, to include with government departments, law enforcement, telecommunications, and finance. She is also a former British Army reservist soldier with 7 years military service, most of which was spent with 4th Battalion, the Parachute Regiment. Dr. Keene has published numerous internal and external MoD and NATO reports as well as award winning academic journal articles. She is the author of "Threat Finance: Disconnecting the Lifeline of Organized Crime and Terrorism." Dr. Shima graduated from the University of Buckingham with honors in business studies and holds an M.Phil. in defence and security studies from the Defence College of Management and Technology, Defence Academy of the United Kingdom, and a Ph.D. in international criminal law from the Institute of Advanced Legal Studies, University of London.

End Notes - Biographical Sketch

[1] In the UK system, DCEs are nongovernment civilians who are available for deployment, often at short notice, for assignments in countries affected by or at risk of violent conflict. They are part of the Civilian Stabilisation Group (CSG) which is a pool of skilled individuals who are deployed to fragile and conflict-affected countries to assist the UK Government in addressing instability.

[2] The Stabilisation Unit (SU) is a civil-military operational unit based in Whitehall, London, which supports UK Government efforts to tackle instability overseas. It is an internationally recognized center of excellence on stabilization, conflict, and security and home of the UK Government's lessons learned on stabilization work. Formed in 2007 as an interdepartmental agency, jointly owned by the Foreign and Commonwealth Office (FCO), Ministry of Defence (MoD), and the Department for International Development, the SU became an independent unit in 2015, funded through the Conflict Stability and Security Fund and governed through the National Security Council.

In: Drone Warfare: Ethical Explorations   ISBN: 978-1-63485-103-9
Editor: Michelle Holloway   © 2016 Nova Science Publishers, Inc.

Chapter 2

THE ETHICS OF DRONE STRIKES: DOES REDUCING THE COST OF CONFLICT ENCOURAGE WAR?[*]

James Igoe Walsh and Marcus Schulzke

FOREWORD

Armed unmanned aerial vehicles—combat drones—have fundamentally altered the ways the United States conducts military operations aimed at countering insurgent and terrorist organizations. Drone technology is on track to becoming an increasingly important part of the country's arsenal, as numerous unmanned systems are in development and will likely enter service in the future. The increasingly frequent use of drones raises profound questions about the nature and morality of warfare involving asymmetrical risks between opposing belligerents. Concerned citizens, academics, journalists, nongovernmental organizations, and policymakers have raised questions about the ethical consequences of drones and issued calls for their military use to be strictly regulated. This level of concern is evidence that the future of drone warfare not only hinges on technical innovations, but also on careful analysis of the moral and political dimensions of war. Regardless of whether drones are effective weapons, it would be difficult to sanction their

---

[*] This is an edited, reformatted and augmented version of a monograph issued by the Strategic Studies Institute, September 2015.

use if they undermine the legitimacy of U.S. military forces or compromise the foundations of democratic government.

One key ethical challenge drones raise is that removing American soldiers from the battlefield could alter civilians' attitudes toward the use of military force in ways that promote war and undermine democratic accountability. Casualty aversion, the civilian public's discomfort with sustaining military casualties and resistance to costly military operations, is a powerful constraint on when and how wars are waged in democratic societies. Political leaders in such polities, and even some high-ranking commanders within the military, may feel pressured by public opinion to wage wars in ways that minimize the risk to soldiers, or to avoid fighting entirely when casualties are likely. One of the most popular and plausible arguments against the use of drones is that these weapons subvert the constraints created by casualty aversion in potentially dangerous ways. Drones may allow wars to be waged without risk to human soldiers and therefore without the risk of provoking public backlash. The weakening of this constraint might permit leaders to initiate or escalate conflicts that, absent the availability of drones, might generate domestic political controversy. In what follows, Dr. Marcus Schulzke and Dr. James Walsh discuss the logic of such arguments against drones, and touch upon counterarguments which suggest that drones might permit war to be waged in a more ethical manner than current weapons technologies permit.

Although the argument that drones will subvert casualty aversion is one of the most common objections raised against these weapons, it has not been subjected to much systematic empirical investigation. It is generally substantiated with inferences drawn from past wars and with theoretical accounts of how drones may promote civic disengagement. The authors assess this argument with a survey experiment. Participants were randomly assigned to read information about fictional conflict scenarios. These scenarios varied the type of attack by U.S. forces, describing it as drone strikes, strikes from manned aircraft, or the use of ground troops. They also systematically altered the strategic goals of the military mission, which included counterterrorism, humanitarian intervention, the restraint of an aggressive foreign power, and support for an ally facing an internal military threat to its hold on power.

Their results show that participants are more willing to support the use of force when it involves drone strikes. Support for attacks increases noticeably when it is described as a drone strike. However, this technology's influence on support for military interventions may not be as profound as critics of drone warfare often argue. Indeed, one important shortcoming of philosophical and ethical reflections on the effects of drones is that they do not produce very precise estimates about how sizable a change in opinion the introduction of this technology will create. An important contribution, then, is to compare how drones alter opinions compared to other factors known from existing research that alter support for the use of force.

For example, gender has an influence on support for war that was comparable to using drones. Existing research suggests that gender has a consistent, but not overwhelming, large effect on attitudes towards military force. While one should use care in generalizing from the results of one experiment, these findings suggest that the possibility of engaging in military action with drones should, in general, increase support for the use of force by a modest amount. The practical consequences of such changes would depend on how closely the public was divided on a proposed military mission.

Many factors are at play when leaders propose to or actually use force. The availability of combat drones may be one such factor, but that alone is unlikely to be decisive in most scenarios. It also indicates that the type of military action matters. Participants were more likely to support wars that posed lower levels of risk to American soldiers, but they were also more likely to support wars in pursuit of important objectives (especially for counterterrorism) and when they thought that war was generally an effective foreign policy tool. This suggests that critics of drones are correct in calling attention to the risk of drones lowering inhibitions against war, but that this shift in attitudes alone is unlikely to have a strong effect on the incidence of wars.

DOUGLAS C. LOVELACE, JR.
Director
Strategic Studies Institute and U.S. Army War College Press

SUMMARY

One of the most compelling arguments that has been raised against drone weapons is that they may lower inhibitions against going to war by making it possible to fight without sustaining casualties. This monograph assesses this argument by using a survey experiment designed to gauge whether American civilians are more willing to initiate wars using unmanned aerial vehicles (UAVs) than using ground forces or piloted aircraft. The use of UAVs made participants more likely to support initiating a war, and this was consistent across four principal policy objectives that were the cause for war:

counterterrorism, humanitarian intervention, foreign policy restraint, and internal political change. However, the increase in support for war caused by UAVs was fairly small, and would probably not be sufficient to tip the balance of public opinion in favor of fighting under most circumstances. Support for war was also heavily influenced by other factors, such as what principal policy objective was being pursued.

INTRODUCTION AND OVERVIEW

Drones have had a revolutionary influence on U.S. military operations over the past 2 decades. This technology is on track to become an increasingly important part of the country's arsenal as the dozens of unmanned systems currently in development enter service in the future. Drones have also raised profound questions about the nature of warfare and the morality of fighting in ways that create asymmetrical risks between opposing belligerents. Concerned citizens, academics, journalists, nongovernmental organizations, and policymakers have spoken out against drones and called for them to be strictly regulated or even prohibited.[1] This level of public concern is evidence that the future of drone warfare not only hinges on technical innovations, but also on careful analysis of the moral and political dimensions of war. Regardless of whether drones are effective weapons, it would be difficult to sanction their use if they undermine the legitimacy of U.S. military forces or compromise the foundations of democratic government.

One key challenge raised by many critics of unmanned aerial vehicles (UAVs) specifically, and unmanned systems more generally, is that removing American soldiers from the battlefield could disrupt civilian attitudes toward the use of military force in ways that promote war and undermine democratic accountability. Casualty aversion, which we understand to be the civilian public's discomfort with sustaining military casualties and resistance against costly military operations, is a powerful constraint on when and how wars are waged in democratic societies. Policy-makers, and even some high-ranking commanders within the military, may feel pressured by public opinion to wage wars in ways that minimize the risk to soldiers or to avoid fighting when casualties are likely. One of the most popular and plausible arguments against the use of drones is that these weapons subvert the constraints created by casualty aversion in dangerous ways. Drones may allow wars to be waged without risk to human soldiers and therefore without the risk of provoking public backlash.

Although the argument that drones will subvert casualty aversion is one of the most common objections raised against these weapons, it has not been subjected to systematic empirical investigation. It is generally substantiated with inferences drawn from past wars and with purely theoretical accounts of how drones may promote civic disengagement. We tested this argument with survey experiments involving over 3,000 participants in the United States recruited from Amazon's Mechanical Turk online labor market. Participants were randomly assigned to read information about fictional conflict scenarios. These scenarios varied the type of attack by U.S. forces, describing it as drone strikes, strikes from manned aircraft, or the use of ground troops. They also systematically altered the strategic goals of the military mission, which included counterterrorism, humanitarian intervention, the restraint of an aggressive foreign power, and foreign policy restraint, and support for an ally facing an internal military threat to its hold on power.

Our results show that participants are more willing to support the use of force when it involves drone strikes. Support for attacks increases noticeably when it is described as a drone strike. However, this technology's influence on support for military interventions may not be as profound as critics of drone warfare often argue. Indeed, one important shortcoming of philosophical and ethical reflections on the effects of drones is that they do not produce very precise estimates about how sizable a change in opinion the introduction of this technology will create. One important contribution of our results, then, is to compare how drones alter opinions compared to other factors that we know from existing research alter support for the use of force. Casualty aversion is one of several considerations that affect support for war, such as mission type and existing attitudes about war. Demographic characteristics like gender, race, income, and age were also included in our analysis, with gender having an influence on support for war that was comparable to using drones. Thus, participants were more likely to support wars that posed lower levels of risk to American soldiers, but they were also more likely to support wars in pursuit of important objectives (especially for counterterrorism), when they thought that war was generally an effective foreign policy tool, or when they were male. This suggests that critics of drones are correct in calling attention to the risk of drones lowering inhibitions against war, but that this shift in attitudes is unlikely to have a strong effect on the incidence of wars.

Our analysis proceeds in five stages. First, we provide an overview of the research on casualty aversion and explore the reasons why low casualty tolerance may limit wars in both *jus ad bellum* and *jus in bello* senses. Second, we discuss arguments that drones may circumvent casualty aversion in ways that lead to an increased incidence of war and undermine democratic accountability. We also raise the possibility that lowering inhibitions against war could have beneficial consequences by making it easier to engage in humanitarian interventions. Third, we explain our research design and show how it improves on aggregate polling data when assessing support for military interventions involving drones. Fourth, we present our results and discuss their implications for the debate over the morality of drone warfare. Finally, we conclude by considering some of the policy implications of our research and call attention to the importance of conducting further research on dimensions of this topic that we were not able to test.

LITERATURE REVIEW

Drones have become the subject of intense debate between those who think that the weapons raise serious ethical challenges that justify their prohibition and those who believe they are ethically advantageous. Much of this debate is focused on armed unmanned aerial vehicles (UAVs), such as the Predator and Reaper, and on the permissibility of targeted killing. However, commentators have also made efforts to develop general theories of drone ethics that account for other kinds of unmanned vehicles that may enter service in the near future. Some have even speculated about the ethical implications of autonomous weapons,[2] though ethical analysis of autonomous weapons is hindered by uncertainty about the meaning of autonomy and whether the military would realistically develop weapons that operate without human control.

Our analysis is primarily directed at the ethical challenges posed by combat UAVs in particular, which is defined as an airframe armed with air-to-ground weapons that is controlled remotely by a pilot who is located outside of the combat zone. There are several reasons for this. First, because UAVs are already being used and play a central role in American military operations, the ethical issues they raise are more urgent than some of those associated with weapons that are not yet in service or that have not been created. Second, given the uncertainty about what form drones will take in the future and what roles they will perform, it is difficult to develop experiments that will reliably gauge public opinion about them. Finally, our findings about UAVs may be generalizable to other types of aerial drones, as well as to drones operating on land and at sea, that have similar capacities for distancing their controllers from the battlefield and protecting them from being attacked. As discussed later, there is good reason to believe that our findings related to UAVs will hold true for other types of drones.

Despite the broad range of issues taken up in the debate over the ethics of drone warfare and the many different perspectives that have been offered, one question has emerged as a central point of contention: will drones lower inhibitions against using military force in future conflicts? Commentators on both sides of the debate over drone use have reflected on this question and given reasons for thinking that drones may lower civilians' inhibitions against fighting, raise new ones, or fail to significantly alter them in one way or another. With limited empirical data on drone usage available, most commentators have sought to substantiate their answers to this question by theorizing the causal mechanisms that might account for drones altering attitudes about the use of military force. Although the causal mechanisms that have been posited in previous research cannot provide a clear answer to questions about the consequences of drone use, they do provide a strong starting point for developing the hypotheses that we will test.

Casualty Aversion as a Constraint on War

By far the most common concern expressed by opponents of drone warfare is that drones may make it easier to use military force by obviating the need for risking human combatants' lives in combat. This line of argument generally starts from the premise that casualty aversion helps to constrain wars *ad bellum* and *in bello*. From a *jus ad bellum* perspective, civilians may anticipate the casualties that could be sustained in a prospective war and oppose the use of military force to avoid the loss of life. Civilian support for an ongoing war might also wane as losses mount, thereby making it difficult to continue fighting in the aftermath of costly actions. From a *jus in bello* perspective, civilians may oppose intensifying wars, extending them geographically, or engaging in certain types of operations when these changes in the conduct of war may result in heavy casualties. Low casualty tolerance may therefore help to limit the occurrence of wars, compel belligerents to seek peace more quickly, and prevent conflict escalation.

There is a great deal of empirical support for the belief that casualty aversion helps to prevent or constrain wars. Multiple studies have found evidence indicating that civilians, especially those who are citizens of liberal democracies, tend to disapprove of wars that result in heavy casualties.[3] Other studies have made the complementary discovery that democratic leaders are vulnerable to drops in public support that can be triggered by costly wars.[4] These findings also hold a great deal of intuitive appeal because they seem to cohere with changes in public support for American military ventures since World War II. For example, low casualty tolerance explains how civilian opposition to U.S. military involvement in Vietnam and Somalia, following costly combat operations like the Tet Offensive and the Battle of Mogadishu, may have precipitated withdrawals from those countries, even when U.S. forces had a significant military advantage over their opponents.

Some question the power of casualty aversion and contend that there is not simply a linear increase of opposition to war as more soldiers are wounded and killed. Gelpi *et al.*[5] argue that many considerations affect casualty tolerance, though the prospect of success seems to be the most important:

> [W]hen it comes to supporting an ongoing military mission in the face of a mounting human toll, expectations of success matter the most. Many factors—the stakes, the costs (both human and financial), the trustworthiness of the administration, the quality of public consensus on the foreign policy goal in question, and so on—affect the robustness of support. But the public's expectation of whether the mission will be successful trumps other considerations.[6]

This suggests that casualty aversion may vary considerably, depending on the type of war being fought.

Others question the reality of casualty aversion and contend that this phenomenon is largely based on false perceptions. Eric Larson finds that:

> [a]s a result of the Gulf War, the public does not expect—and is unlikely to demand—that all future U.S. military operations be bloodless. Indeed, it is more accurate to say that the public hopes for low-to-no casualty operations but fears a very different outcome.[7]

Similarly, Charles Hyde argues that casualty aversion is a myth and says that "[t]he public has consistently operated within the realm of an ends and means evaluation with significant cues from political leaders who frame the public debate."[8] Casualty aversion may be closely linked to media coverage that focuses disproportionately on anti-war sentiments when soldiers are killed, as this kind of coverage may erode public support or create the false appearance of a drop in support even when real attitudes are fairly stable.[9]

Regardless of whether it is constant or variable, real or imagined, casualty tolerance does seem to be a consideration that civilian politicians and military leaders bear in mind when considering how and when to use military force.[10] Even Hyde, who describes casualty aversion as a myth, finds that there is "strong evidence that policymakers and senior military leaders believe the American public is casualty averse and will not tolerate deaths except when vital interests are at stake."[11] Low casualty tolerance may therefore be expected to create *ad bellum* and *in bello* inhibitions that limit the incidence of wars and their intensity either because of direct pressure from the public or because of a perceived risk of public opposition. In either case, the resultant pressure may compel politicians and military leaders to seek nonmilitary strategies of conflict resolution or to at least use the utmost restraint when fighting.

Drones and the Future of Risk-Free War

To their critics, drones seem to circumvent the restrictions of casualty aversion. They make it safer for armed forces to wage wars without exposing their soldiers to the hazards of the battlefield, thereby mitigating the possibility that casualty aversion, whether real or perceived, will influence the decisions of politicians and military leaders. Moreover, critics worry that if drones allow the U.S. military to wage wars free of the risk of military casualties, they will further shift the burdens of war onto foreign populations. This may result in more wars, more foreign casualties, more environmental destruction, and more social and political disruption in contested areas, all without provoking much domestic opposition.

John Kaag and Sarah Kreps argue that "drones create a 'moral hazard' by shielding US citizens, politicians, and soldiers from the risks associated with targeted killings."[12] Without risk, civilians and soldiers alike may be unable to understand the consequences of war. This may lead them to see war as a cheap and effective solution to complex political disputes that could be more effectively solved through peaceful means. The ethical principles that inform judgments about war may even lose their power when these are weighed against the compelling practical advantages of using drones. Kaag and Kreps suggest that this freedom from consequences may lower the threshold for initiating wars and make it easier to sustain protracted wars that might otherwise be forced to a conclusion by declines in public support. They are also concerned that drones may make it easier to wage wars in ways that are ethically and legally questionable, as evidenced by the use of UAVs to carry out targeted killings in countries with which the United States is not at war.

By Kaag and Kreps' estimation, the moral hazard of drones is not merely a possibility, but a likely outcome given the incentive structure that shapes politicians' decisions. Drones are attractive to politicians who may see war as a way of achieving their foreign policy goals but whose power is threatened by any shifts in public opinion that a costly war might cause.

> The use of drones provides a win–win proposition for the president, who could appear strong on defense without responsibility for body bags coming home, a development that would likely send his political fortunes tumbling.[13]

Kaag and Kreps contend that politicians are likely to increase their reliance on drone weapons over time, since those who fail to do so will be met with public opposition and risk losing office.

Even more seriously, Kaag and Kreps think that the incentive structure drones create could ultimately have deleterious effects on the U.S. Government by making it possible for politicians to wage wars without securing public approval:

> Ironically, the pressure from a democratic electorate to protect itself from the harms of warfare will not encourage policy makers to adopt peaceful or democratic methods . . . but rather methods of warfare that leverage technology in order to insulate citizen-soldiers from harm. The irony is this insulation creates the possibility that leaders will no longer, in a prudential sense, have to obtain popular permission to go to war.[14]

The threat of de-democratization exacerbates the potential problems associated with drone use and suggests that even Americans who are unconcerned with the effects U.S. military operations have abroad should be alert to the possibility that overcoming casualty aversion could erode democratic governance.

Kaag and Kreps' concern over the loss of democratic accountability builds on a point previously made by P. W. Singer, who describes drones as being part of a larger process of de-democratizing war as armed forces increasingly fight in ways that evade public accountability.

Unmanned systems represent the ultimate break between the public and its military. With no draft, no need for congressional approval (the last formal declaration of war was in 1941), no tax or war bonds, and now the knowledge that the Americans at risk are mainly just American machines, the already lowering bars to war may well hit the ground. A leader needn't carry out the kind of consensus building that is normally needed before a war, and doesn't even need to unite the country behind the effort.[15]

Singer goes on to describe a collapse of civic engagement as the low costs of war lead decisions about the use of military force to become routine policy decisions that provoke little serious public deliberation. By his reasoning, drones will reduce war to a spectator activity as "the checks and balances that undergird democracy go by the wayside."[16] Thus, like Kaag and Kreps, Singer sees casualty aversion as a powerful mechanism for limiting war that has implications for reducing aggression and sustaining accountability.

Frank Sauer and Niklas Schörnig[17] contend that drones are attractive weapons for democratic states for procedural and normative reasons. Procedurally, democratic leaders can be punished for waging wars that result in heavy casualties by being removed from office. This compels them to seek the least costly methods of conflict resolution that are available and to avoid war whenever possible. Normatively, democracies place a high value on human life, especially the lives of their own citizens, which provides added incentive to avoid violence. Sauer and Schörnig argue that these concurrent influences usually compel democracies to act peacefully, but that democracies may escape these restrictions when they are able to wage wars either without sustaining casualties or without **appearing** to do so.

Many techniques are available to democracies attempting to minimize the costs of war. They can employ special operations forces or private military contractors that may suffer fewer or less visible casualties. They can also launch guided missiles, which have the advantage of not endangering human combatants on the attacker's side. However, Sauer and Schörnig describe drones as the "preferred solution" to the casualty tolerance problem because drones are relatively cheap, do not put the military personnel operating them at risk, and are militarily effective. Sauer and Schörnig share Kaag and Kreps' concern that drones will make it easier for states to engage in risk-free warfare and to evade accountability. They also contend that this problem is apt to follow a slippery slope. As drones make wars safer for the militaries armed with them, they will become more pervasive and more autonomous, which will, in turn, lead to even lower inhibitions against using force and increased demand for drones.

Daniel Brunstetter and Megan Braun[18] speculate that drones could have the paradoxical effect of simultaneously making major wars less likely and small wars more likely. Their reasoning is that drones alter the considerations affecting the *jus ad bellum* principle of last resort, which requires that states pursue all available peaceful means of conflict resolution before employing military force. Drones help to prevent large wars by giving states a capacity to carry out small strikes that are less likely to provoke escalations than combat between human soldiers. However, Brunstetter and Braun contend that drones may lead states to deviate from the principle of last resort when power asymmetries make it possible to carry out drone attacks without fear of reprisal. As they explain:

> [t]he risk becomes that military leaders will bypass nonlethal alternatives, such as apprehending alleged terrorists and continued surveillance, and move straight to extrajudicial killing as the standard way of dealing with the perceived threat of terrorism.[19]

The perception that drones are more discriminate than other weapons exacerbates this problem by making it possible to present drone strikes as a form of controlled violence that is less serious than war.

Many other writers also express concerns about drones' effects on civilians' attitudes about war, though usually without going into as much detail about the exact mechanisms underlying this process. Linda Johansson argues that:

> [w]eapons such as UAVs, that are perceived to provide an almost guaranteed upper hand, may also make people believe that the war would be without risk, at least regarding the number of casualties

and concludes that "[t]his might have an impact on domestic opinion, and, in turn, lower the threshold of entering and sustaining a war."[20] Christian Enemark[21] likewise expresses concern over what war may be like in the future as more states and nonstate actors develop unmanned weapons.

> The prospect of increased availability of armed drones warrants contemplation of a future in which the resort to force is less constrained

by the expectation of loss, and this sits uneasily with ethical principles that have traditionally set a high threshold for going to war.[22]

Enemark argues that drones have already demonstrated this capacity for lowering inhibitions against war and that this is evidenced by the U.S. preemptive war against potential terrorist threats.[23]

David Dunn[24] reflects on the way drones have altered casualty calculations in the War on Terror, with the effect of incentivizing more lethal strategies.

> By disembodying these weapons platforms, the technology enables their use with domestic political impunity, minimal international response and low political risk and cost. It is now politically and technically easier to kill suspected terrorists than to arrest them.[25]

Boyle[26] raises a similar point by suggesting that superpowers that were once afraid to fight each other for fear of triggering nuclear war might suddenly find it easier to come into conflict using drones. He reasons that drones could cause subtle provocations, such as reconnaissance missions and small attacks, that would risk escalating into more serious confrontations. Finally, Gurcan[27] argues that drones may make deterrence more difficult, as aggressors armed with drones may not be easy to intimidate if they can fight without sustaining casualties. This reasoning indirectly relies on assumptions about casualty aversion, since one of the goals of deterrence may be to convince civilian populations that a prospective war will be too costly to be worthwhile.

Can Risk-Free War be Ethically Advantageous?

Many commentators who defend drones agree with critics in thinking that drones may circumvent casualty aversion, yet they draw much different conclusions from this premise. Rather than seeing the decline in military casualties as a mechanism for overcoming inhibitions against the use of military force, defenders of drones tend to think that lower casualty rates will make it easier to promote compliance with the norms of just war. Zach Beauchamp and Julian Savulescu[28] give two reasons for thinking that lowering the threshold for initiating wars may be a positive development. First, this may make states more inclined to fight humanitarian conflicts. If states do not have to risk their own forces, they will be free to wage benevolent wars that do not yield strategic benefits without facing public backlash. Second, Beauchamp and Savulescu argue that being freed from the fear of sustaining casualties will allow intervening states to show higher levels of restraint when fighting. This lends additional support to the use of drones in humanitarian wars, as it suggests that drones may be used in ways that coincide with the values of human security and respect for civilian immunity, which help to justify humanitarian missions.

The points raised by Beauchamp and Savulescu, particularly the second, coincide with those made by others who think that drones can be ethically advantageous. Strawser[29] argues that armed forces have an obligation, which he calls the Principle of Unnecessary Risk, to prevent their personnel from being exposed to avoidable risks. By this account, it would be unethical for armed forces that have drone technology to fail to use this technology as a way of protecting soldiers. Moreover, Bradley Strawser and others call attention to the benefits this may have for civilians. Freeing armed forces from the concern over sustaining casualties could allow them to establish much stricter rules of engagement that might help to reduce violence against civilians.[30] Drones do not face the same need to act in self-defense as human combatants do. Their operators could be prohibited from using lethal force whenever there is a high risk of inadvertently harming civilians without raising any corresponding danger to military personnel. Drones might therefore be a way of escaping the ethical dilemma of whether to prioritize force protection or civilian protection, which has hitherto presented an insurmountable challenge for just war theorists.[31]

These arguments in defense of drones show that any effect these weapons have on casualty aversion can be read in much different ways, depending on whether casualty aversion is seen as a constraint on war or a way of facilitating the just conduct of wars. This makes it vital for analyses of how drones influence casualty tolerance to account for the different types of wars that drones may be used in, as well as whether drones are used in ways that increase or decrease the risks to civilians. We do this in our experiments by testing support for drone strikes, air strikes, and ground attacks in pursuit of four different principal policy objectives: foreign policy restraint, counterterrorism, humanitarian intervention, and internal political change. Before turning to these experiments, we first discuss how some of the points raised earlier are reflected in public opinion surveys regarding the conflict with the Islamic State armed group in 2014. The debate about American military intervention against the Islamic State included discussions not only of the wisdom of intervening, but also the merits of different types of military action. It thus provides a contemporary opportunity to assess how the availability of drone technology influences public attitudes.

Public Opinion and the Islamic State

The United States experienced public debates about the wisdom of American action against Islamic State militants in Iraq and Syria during the summer and fall of 2014. Much of this debate centered on the type of military action, if any, the United States should undertake. A number of public opinion organizations polled representative samples of the American public and asked the degree to which they favored or opposed a range of steps being considered by the United States.

Consider the data in Figure 1, which summarizes responses to questions about favoring or opposing different types of intervention in the conflict. A sizable majority of respondents favored air strikes in Syria and in Iraq, while far fewer supported the introduction of American ground troops. This is consistent with the argument that technologies that reduce the costs of conflict by placing fewer military personnel at risk of harm, such as drones and air power, should lead to increased support for the use of force. Note, however, that the option of sending military advisors to Iraq receives almost as much support as does the use of air power. This is a bit puzzling from this perspective, as military advisors may be in proximity to Iraqi ground troops who engage in combat. Note as well that training and equipping rebels in Syria, which presumably would not risk combat by Americans, receives considerably less support. Why might this be the case? One explanation might be that the public also incorporates its beliefs about the likelihood that the use of force will succeed in achieving its goals against its human and financial costs. It is possible that respondents recognized that sending military advisors to Iraq would place them in harm's way, but balanced this against the belief that advisors could bolster the effectiveness of Iraqi ground units against the rebels. They may also have concluded that arming Syrian rebels, who had proven incapable of either overthrowing the Assad regime or stopping the rise of the Islamic State, would be an ineffective strategy.

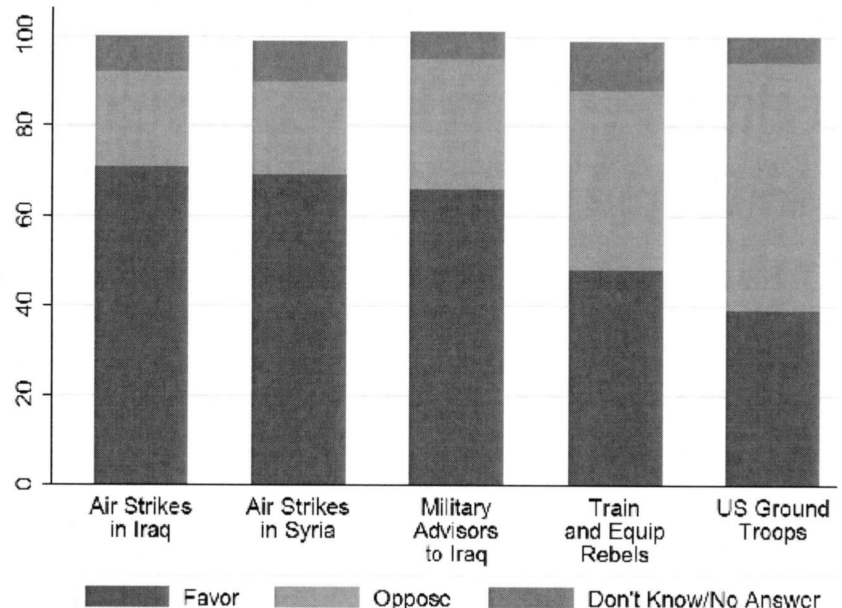

Notes: Data from CBS/*The New York Times* public opinion survey released September 17, 2014. Totals do not equal 100 due to rounding.

Figure 1. Support for Military Action, September 2014.

Some evidence for this comes from comparing the information in Figures 2 and 3, which summarize responses to questions about preferences for different combinations of air power. Figure 2 depicts responses to questions that asked respondents if they favored the use of drones and manned aircraft in striking the Islamic State. Drone strikes received considerably more support than attacks from manned aircraft, which is consistent with the argument that reducing the costs of war increases support for the use of force. From this perspective, though, the results of a poll conducted at approximately the same time are surprising. As can be seen in Figure 3, a much higher percentage of respondents prefer strikes from platforms that do not risk American lives—drones and cruise missiles— than from piloted aircraft. However, an even larger percentage prefer strikes from **both** types of weapons systems. This might be because such combined strikes are seen as more effective than the use of only one weapons system.

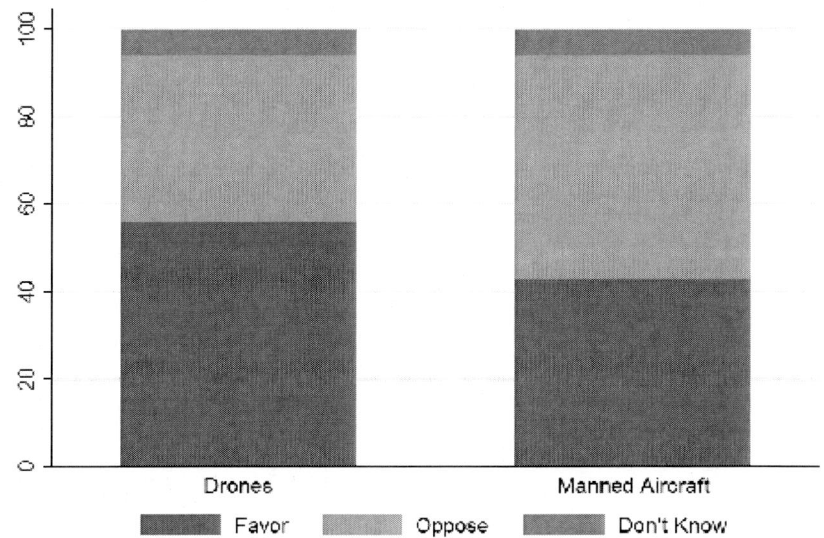

Notes: Data from CBS/*The New York Times* public opinion survey released June 23, 2014.

Figure 2. Preferences for Strikes from Manned Aircraft and Drones, June 2014.

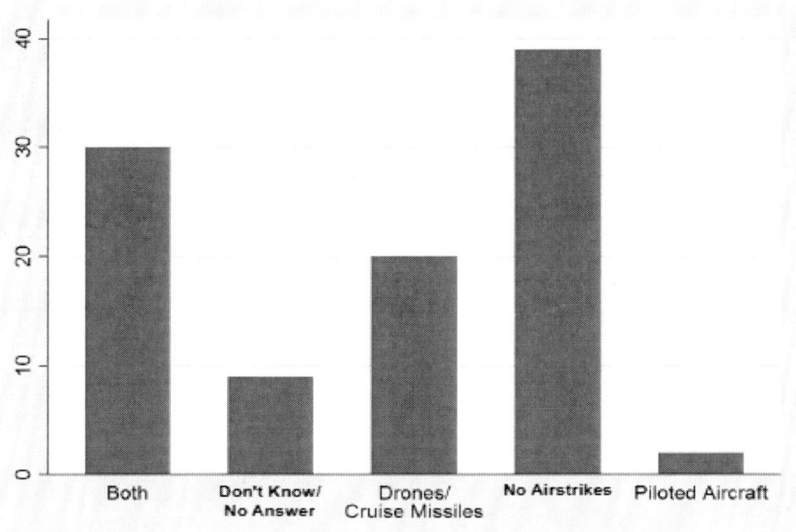

Notes: Data from Quinnipiac public opinion survey released July 3, 2014.

Figure 3. Preferences for Strikes from Manned Aircraft and Drones, June 2014.

This public opinion data, then, does not reveal an obvious pattern in which attacks that reduce the likelihood of an important cost of war—military casualties—receive more public support. It is important to recognize, though, that such polls are not really designed to directly assess specific propositions about how changes in the costs of war influence attitudes. To see why this is the case, return to Figure 1. We suggested that one reason respondents in this survey expressed strong preferences for sending military advisors to Iraq was that they believed that doing so would increase the combat effectiveness of Iraqi units while exposing American military personnel to acceptably small risks of harm. But other interpretations are plausible; for example, perhaps many respondents believe that advisors would face very small risks roughly equal to those of aircraft crews launching attacks in uncontested air space. The questions in this and most public opinion surveys are intended to measure support for various options, not to assess directly why respondents express the preferences that they do. Yet doing so is important for establishing which of the propositions about how changes in the cost of conflict influence public support for the use of force. In the next section, we argue that an experimental approach is better-suited for this purpose, and summarize results from a series of experiments that are designed to directly test such propositions.

Survey Experiment

To answer these types of questions, we conducted a survey experiment in early 2015. This experiment recruited participants from Amazon's Mechanical Turk online labor market. Mechanical Turk is an inexpensive and flexible way to enlist participants, and has become a widely used tool among social scientists.[32] Respondents were randomly assigned to read a mock news story describing plans by the United States to use military force overseas.[33] Random assignment to these "treatments" is a key part of the experiment. It allows us to assume that the characteristics of individuals assigned to read each news story are not systematically different from each other. This means that we can expect that any differences in the attitudes that people assigned to different treatments are due to the content of the news stories.[34]

The news stories varied two elements (see the Appendix for the complete wording of each treatment as well as the questions that comprise the survey instrument). The first was the type of military action. This could take one of three forms: a drone strike, an air strike from a piloted aircraft, or the use of ground troops. Consistent with the casualty aversion idea discussed earlier, the news stories had different information about the risk that American military personnel would face. The drone treatments stated that "the use of unmanned drones means that no American military personnel would be placed at risk." The air strike treatments, in contrast, included information that the target of the strikes were believed to lack weapons capable of attacking aircraft, suggesting a low possibility of military casualties. The ground troops news stories did not mention if these troops faced any danger or not.

The second element that varied across the treatments was the purpose or goal of the use of force. Here, we follow in the footsteps of important work on public opinion and foreign policy which finds that preferences differ in important ways depending on the "principle policy objective" of the use of force.[35] The treatments in our news story vary four such objectives. The first is **counterterrorism**, in which attacks are planned on militants who have in the past attacked the United States. The second is **foreign policy restraint**, where the United States seeks to punish a foreign state for threatening a key interest, in this case the shipment of petroleum from the Persian Gulf to world markets. The third is **humanitarian intervention**, in which American military force has the objective of stopping mass killings in a foreign country. The final objective is **internal political change**, aimed at preventing the violent overthrow of a foreign government by its internal opponents. Our mock news stories are modeled closely on those of Gelpi, Feaver, and Reifler[36] and use the country of Yemen as the location of the use of force.

Combining these two elements—type and objective of military force—produces a total of 12 treatments. Roughly 300 participants were randomly assigned to read each of these stories. They then answered questions about their reactions to the planned use of force, including the degree to which they supported the attack; their estimates of the number of military casualties that would result if the attack were carried out; general attitudes regarding the wisdom of the use of force; and demographic questions such as party iden-tification, age, gender, and so on.

To this point, we have argued that military casualties are a very important cost of conflict that influences attitudes regarding the wisdom of using force. Our experiment is designed to alter systematically the likelihood of such casualties. We expect that participants in the experiment will expect the lowest number of military casualties from a drone strike, since these news stories make explicit the fact that military personnel will face no risk of physical harm. Treatments involving air strikes should lead to higher expectations of military casualties. Even though these treatments state that the target of the attack is not believed to have weapons capable of threatening military aircraft, participants might still expect that the chance of military casualties could be higher since such aircraft do place military personnel in a battle zone. Participants might worry that the target has, unknown to the United States, acquired anti-aircraft weapons, or that casualties could result if the aircraft were to malfunction over enemy territory. Participants' expectations of military casualties should be highest in the treatments that describe the attack as being carried out by ground troops. Although these news stories make no mention of the risks that military personnel face in such situations, it should be straightforward for participants to infer such risks from the information they read.

To assess how assignment to different treatments influenced estimates of military casualties, participants were asked if they expected no casualties, between 1 and 10 casualties, or between 11 and 100 casualties. The black dots in Figure 4 display the average responses to these questions for each treatment; the lines connected to each dot indicate the 95 percent confidence interval surrounding these averages. We see a pattern, consistent with our expectations, in which participants assigned to the drone strike treatments expected the fewest casualties, followed by those assigned to air strike treatments and then the ground troop treatments. Differences across treatments for the same mission objective are sizable and statistically significant. Furthermore, differences across treatments for the same type of attack are not statistically different from each other. This suggests that the information in the treatments influenced participants' expectations of the costs of conflict in terms of military casualties.

But do such differences matter for support for the use of force? To answer this question, we asked participants to indicate the degree to which they supported the military action described in the news story they read. Participants could strongly disapprove, somewhat disapprove, somewhat approve, or strongly approve of the attack. The average responses (and confidence intervals) for this question are depicted in Figure 5. Drones lower inhibitions against initiating armed conflicts, as many critics of this technology have predicted. Respondents were consistently more likely to favor the use of UAVs over ground forces in each of the experiments, regardless of the objectives being pursued. They were also more willing to initiate conflicts using drones than piloted aircraft, except in humanitarian interventions. Furthermore, the consistent preference for air strikes over attacks involving ground forces provides evidence that support for a prospective operation generally increases as the likelihood of sustaining military casualties decreases. This indicates that American civilians are more inclined to support using weapons that reduce the risk of military casualties, regardless of whether UAVs are available, and that other weapons that allow U.S. forces to manage risks may produce similar shifts in support for launching an attack.

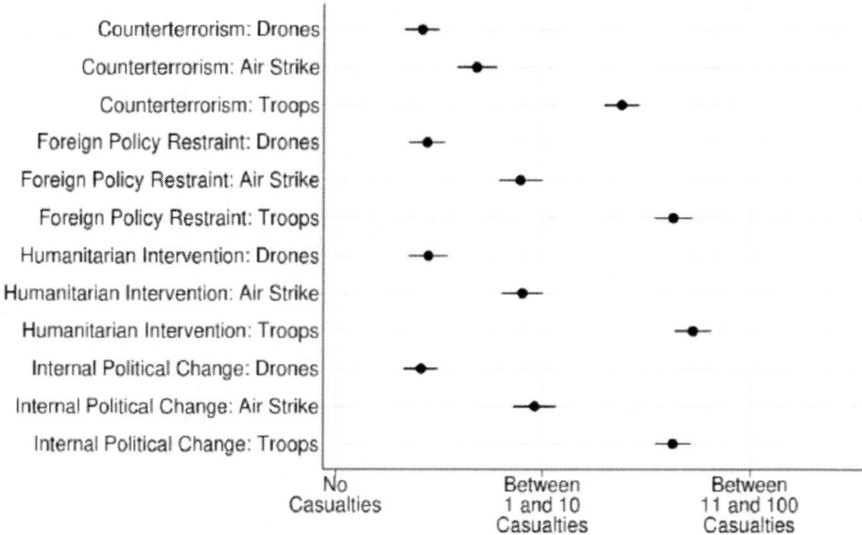

Figure 4. Expectation of Military Casualties.

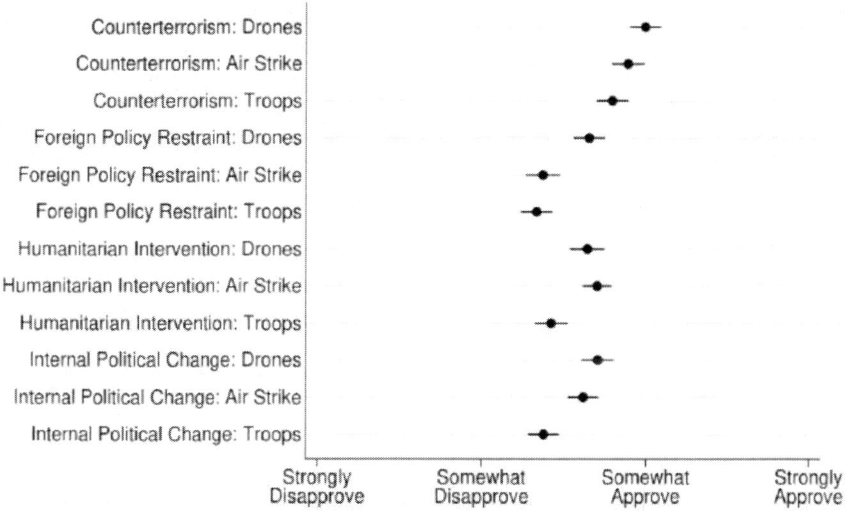

Figure 5. Support for the Use of Force.

Nevertheless, the differences between levels of support for the three tactics are rather small. Respondents were more likely to favor attacks involving UAVs over attacks involving piloted aircraft or ground forces, but on average differences in these treatments reduced the average degree of support from just below "somewhat approve" towards "somewhat disapprove." These small absolute changes in public support for the three tactics suggests that, although UAVs may be expected to lower inhibitions against initiating wars by shielding American soldiers from risk, they are unlikely to increase dramatically the incidence of fighting as some critics of drone warfare have suggested. This is evidence that drones do raise legitimate concerns about attitudes toward initiating hostilities, but that these concerns must be stated far more modestly and that greater attention must be given to how support for drone use varies depending on the context.

Choosing to use drones instead of piloted aircraft or ground troops may be expected to provide a small increase in public support for a prospective war. Even a small change could have a decisive influence on the overall level of public support for war if the country is narrowly divided. In other words, deciding to use drones to carry out an attack could tip the balance of a nearly even division between pro-war and anti-war attitudes toward the former position. On the other hand, if opposition to war outweighs support, then it appears that the use of drones to fight without the risk of incurring casualties would be insufficient to sway public opinion to support an attack. The slight influence on attitudes toward the initiation of hostilities suggests that although drones may diffuse concerns about sustaining casualties, they are unlikely to have a significant effect on the incidence of wars. It seems that unpopular wars will remain unpopular even if drones are able to reduce the level of opposition to fighting.

Another way to assess the substantive influence of drones is to compare their effect to other factors that influence support for the use of force. Figures 6 to 9 undertake such comparisons. Each figure reports the coefficients and associated confidence intervals for a regression model using support for the use of force as the dependent variable. The independent variables in each model include whether or not the participant was assigned to read a story about a drone strike, whether or not the participant read about an air strike, and the party identification, gender, race, income, and age of the participant. All of these variables were rescaled to range from zero to 1. This means that we can compare the coefficients directly to each other. The dots indicate how increasing the independent variable from its minimum to its maximum value influences support for the use of force. For example, the coefficient for drone strike indicates how much support changes when the participant reads about the use of a UAV compared to the use of ground troops, while the coefficient for air strike indicates the change in support when the participant is assigned to a treatment describing such an attack. Similarly, the coefficient for the female variable indicates the change in support when the participant is a woman compared to a male participant. The figure for party identification indicates the difference between participants that self-identify as "strong Democrat" and "strong Republican."[37]

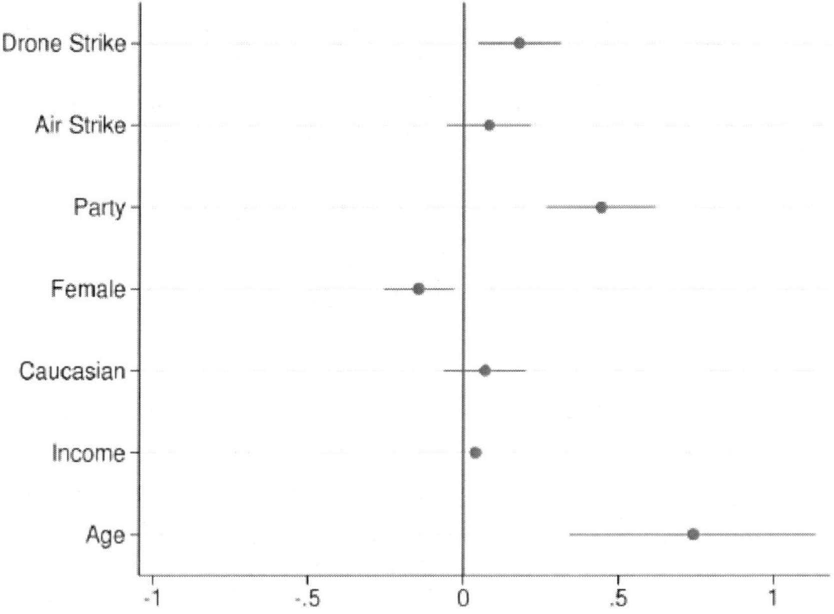

Figure 6. Coefficients for Counterterrorism Treatments.

We see that the effect of describing an attack as a drone strike is roughly similar in magnitude to the effect of gender, and smaller or about the same size as party identification. Both of these factors have well-established influences on support for the use of force and attitudes towards public policy more generally. Pamela Conover and Virginia Sapiro[38] find that gender is one of the most consistent and powerful influences on attitudes regarding military action.[39] Similarly, a large number of studies have concluded that party identification has been shown to have a large effect on attitudes towards many types of public policy.[40] The fact that the substantive influence of drone strikes, compared to ground troops, is of a similar size to the effects of gender and party identification indicates that this weapon could have an effect on attitudes that matters at the margin.

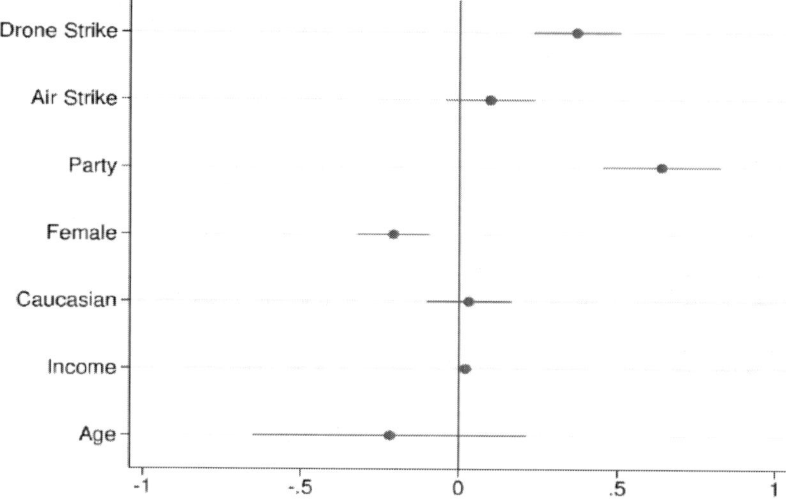

Figure 7. Coefficients for Foreign Policy Restraint Treatments.

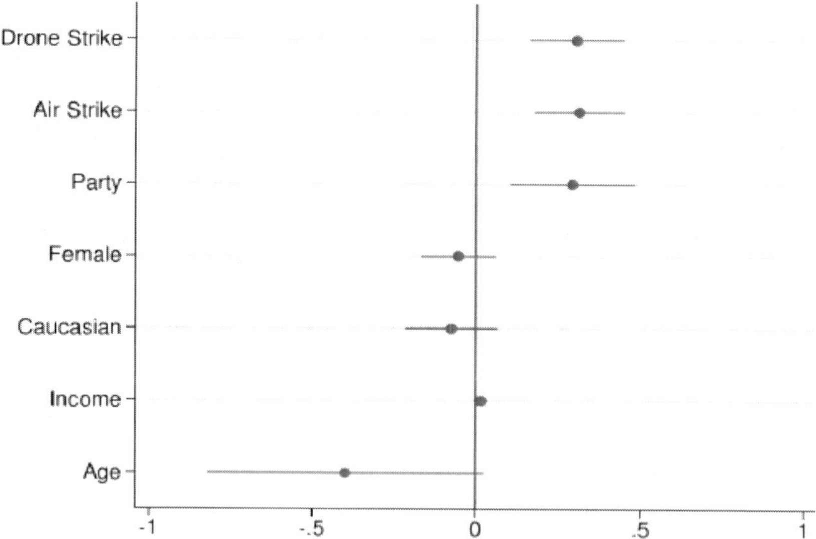

Figure 8. Coefficients for Humanitarian Intervention Treatments.

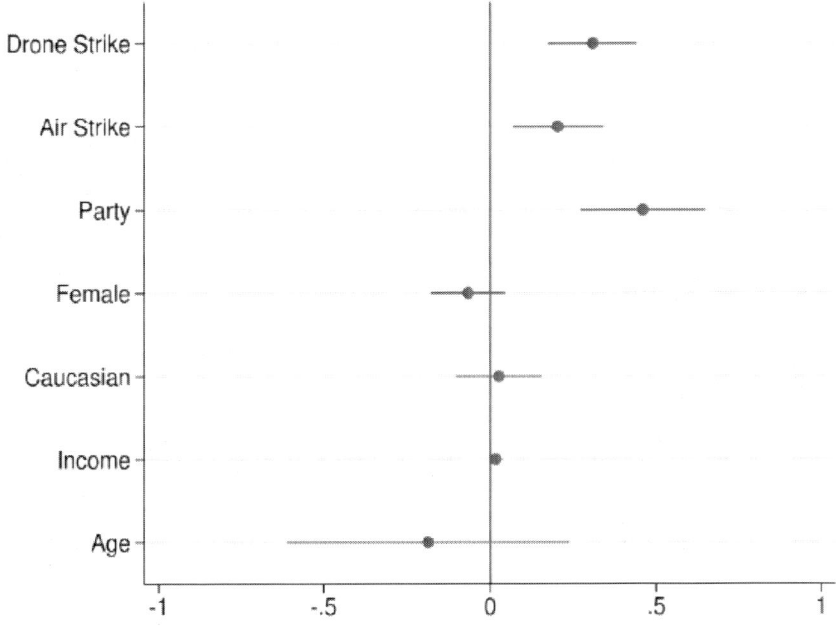

Figure 9. Coefficients for Internal Political Change Treatments.

In practice, the effects of the slight shifts in public opinion that may be produced by using ground forces, piloted aircraft, or drones may not directly affect whether a particular war is waged. Decisions about initiating wars are not made directly by the American public but rather by elected officials whose decisions may be insulated from citizens' attitudes or unresponsive to them. Thus, public opinion will matter to the extent that changes in it can influence policymakers and alter their decisions. Voters may decide to punish policymakers who wage unpopular wars by removing them from office.[41] However, it seems unlikely that the shifts in public opinion that may be caused by using UAVs instead of ground forces or piloted aircraft will always have great influence on policymakers' decisions if they are determined to go to war. An increase in opposition may not greatly influence a policy- maker's election prospects, especially when it is weighed against other decisions that person has made and that will also influence voter preferences.

The effect UAVs have on support for war is likely to make them attractive weapons for politicians who are concerned with maintaining their approval ratings during conflicts. The shift in public support that drones produce may not be enormous and may not be sufficient to cause a decisive change in the balance of public opinion about a war, yet it is nevertheless just one of the many advantages that make drones attractive weapons. As Sauer and Schörnig correctly point out, there are multiple reasons for preferring drones over other weapons and tactics, such as their ability to loiter over targets and their comparatively low cost.[42] Nevertheless, based on our experiment, it appears that the predictions Singer[43] and Kaag and Kreps[44] make about drones undermining democratic accountability are probably too strong. Because drones produce moderate increases in support for war, greater reliance on them will probably be unable to silence antiwar voices. At least some politicians will continue to have strong incentives to pursue peaceful strategies of conflict resolution in an effort to satisfy those citizens who are unwilling to support military operations even if they exact a low human cost.

The pattern that is evident in the levels of support for using drones, piloted aircraft, and ground forces shows that there is a continuum of support for military force that extends across the range of weapons and tactics that may be employed. One possibility for future research is to include a more diverse assessment of the weapons and tactics used by the U.S. military to see the extent to which this pattern is sustained. There may be gradations of support between the three types of attacks we explore. For example, support for the use of special operations forces could fall somewhere between support for an attack involving conventional ground forces and an air strike. Alternatively, it is possible that support for war could further diminish or increase as other weapons and tactics are introduced. The deployment of large numbers of soldiers to directly engage in combat could be less popular than the deployment of smaller numbers of soldiers serving in advisory roles.

Nevertheless, it seems that our study has reached a limit when it comes to how drones may increase support for war by circumventing casualty aversion. The UAVs that were discussed in our experiments provide their pilots with complete protection against harm, as the pilots are removed from the battlefield and cannot be injured by any attacks on their aircraft. Other types of drones, such as semi-autonomous and autonomous drones, may reduce human involve- ment in attacks and alter public support for war, yet they will not do this by influencing casualty calculations. After all, more advanced drones will not be able to offer additional protection for American personnel who are already far away from the battlefield. This indicates that other types of drones may produce slight change in public support for using military force that are comparable to the shifts produced by UAVs.

THE IMPORTANCE OF CONFLICT TYPE

The principal policy objective was an important predictor of participants' willingness to conduct military operations. Participants were most likely to support the use of ground troops, piloted aircraft, or drones in counterterrorism operations, a finding consistent with earlier research.[45] The willingness to support the use of force against terrorists seems to reflect the perceived threat that terrorism poses to national security, as well as the sense that the United States is actively engaged in a War on Terror, in which terrorists are legitimate military targets. Higher support for using drones against terrorists demonstrates the importance of distinguishing between objections to how UAVs are used to conduct targeted killings against terrorists and how they are used in other contexts. The higher levels of support for all three tactics in counterterrorism strikes is also evidence that many of the trends in U.S. military operations that critics have cited as being byproducts of drone use may be more accurately described as being byproducts of the War on Terror that may change if the U.S. military's objectives change.

The other three principal policy objectives involved attacks on less immediate threats against enemies that the United States was not actively fighting when the experiments were conducted. In these instances, fewer respondents were willing to support a war, even if it could be waged with minimal risk to American soldiers. This supports previous research by Bruce Jentleson,[46] who found that Americans were more likely to support attacks when an "adversary had gone beyond simply posing a standing threat and initiated aggressive actions against American interests or citizens" and that the public was not likely to support military operations that were directed at preventing future threats or remaking foreign governments.

The effect of conflict type on support for war suggests that the perceived necessity and morality of war matter to the American public, with more serious threats lowering inhibitions against using any type of force even as considerations about the expected number of American military casualties help to determine how an attack should be carried out. The implication here is that the use of drones is more likely to contribute to the case for initiating war when policymakers can connect drone strikes to plausible enemy threats. The framing of attacks will therefore help to determine whether drones will contribute to a greater incidence of wars.

The varying levels of support for attacking under differing circumstances provide additional evidence that the effects of circumventing casualty aversion with weapons that reduce the likelihood of casualties will be fairly modest.

Avoiding American military casualties is just one of the considerations that go into deciding whether to fight, and it may not even be the most salient one. It is also important to note that, despite the differing levels of support for each type of conflict, participants' attitudes toward the use of military force were generally moderate. The average responses consistently fell between the "somewhat approve" and "somewhat disapprove" evaluations. The participants' general lack of strong attitudes for or against war indicates that participants took a pragmatic attitude toward use of force decisions, according to which attacks are supported or opposed as considerations relating to their necessity, prospective costs, and ethicality change.

Perhaps the most surprising discovery was that support for air strikes in humanitarian interventions surpassed support for UAV strikes. This was the one instance in which the results deviated from the pattern of decreasing levels of support as military personnel were at a higher risk of being killed or injured. It is difficult to determine why this anomaly exists given the available information, especially since this pattern deviates so clearly from the results relating to other principal policy objectives. Because the pattern was present in experiments involving participants with high and low military assertiveness, this preference does not appear to be affected by militaristic attitudes. The different pattern here may be an indication that respondents were evaluating humanitarian interventions according to different criteria than those they applied to other principal policy objectives.

What is clear is that the greater support for air strikes, rather than drones, in humanitarian interventions conflicts with Beauchamp and Savulescu's[47] contention that drones might lower casualty aversion in morally advantageous ways. They argue that drones could promote support for humanitarian interventions by reducing the costs those operations may have on armed forces that conduct them for benevolent motives and without any expectation of compensation. Because support for humanitarian intervention is highest when piloted aircraft are used, it appears that these would be the optimal weapons to employ when intervening and that drones' effects on casualty aversion may only be significant when the only alternative is the more costly use of ground forces. Of course, this may not be true for the many other countries that are developing drone weapons and that may use them in humanitarian missions. It is therefore important to withhold judgment on whether Beauchamp and Savulescu's argument is accurate in general and to continue testing it in future research.

CONCLUSION

Our experiment has important implications for members of the military and policymakers. Although the American military plays only an advisory role in decisions to initiate wars, it unavoidably affects policymakers' calculations about the use of military force and is, in turn, affected by the choices made by policymakers and the American public. Policymakers may have the ultimate control over when wars are declared, but political calculations are affected by the military's decisions to develop certain offensive capacities, as well as its ability to realize those capacities in practice. Members of the military have a large stake in decisions about the resort to war, as they will be the ones who bear the greatest burdens during a conflict. Our results show that drones are unlikely to dramatically change calculations about initiating war in ways that would increase the incidence of fighting, yet the noticeable shifts in public opinion when using drones compared to other weapons reveals that the possibility of drones contributing to the overall case for military interventions cannot be disregarded.

The U.S. military has responsibilities to its soldiers, civilian policymakers, and the American public that could potentially come into conflict when using drone weapons. In a sense, the concern that drones could lower inhibitions against the use of force is a concern that politicians or members of the military may violate public trust by waging unnecessary or aggressive wars. The military could inadvertently fail in its responsibility to protect the American public if it develops weapons that ultimately increase the prevalence of war, especially if wars are economically costly or increase the likelihood of future attacks against the United States. At the same time, the military has an obligation to protect its personnel from the dangers of the battlefield to the greatest extent possible, which will inevitably provide grounds for making greater use of drones and other remote weapons.

Our results demonstrate that these obligations are not perfectly aligned; there is some tension between responsibilities toward American civilians and responsibilities toward American soldiers. Relying on drones to help protect soldiers from harm generally increases support for war in ways that could influence on the decision to use military force when public opinion is narrowly divided and politicians are highly sensitive to it. Although this is a fairly remote possibility given the complex assortment of motives that affect civilians' attitudes and policymakers' insulation from public pressure when initiating wars, it is an important possibility for members of the military to bear in mind. Decisions to expand the U.S. military's drone force and to employ this technology in a greater range of combat roles should be made with some sense of how developing certain technical capacities might alter future decisions to initiate wars.

Although our experiment focused on the UAVs that are currently used by the U.S. Air Force and Central Intelligence Agency, the results speak to concerns that will affect the Army as it develops new unmanned weapon systems and prepares to deploy them in future conflicts. The Army has consistently worked towards improving its ethical standards since the Vietnam War, as evidenced by its continual reevaluations of ethics education, its efforts to promote core values, and the Center for the Army Profession and Ethics' work developing more effective ethical training tools.[48] As the Army strives to protect its values and promote ethical conduct in the future, it will be important for it to remain aware of the importance of ensuring that its efforts to improve force protection do not create new ethical challenges. Most of all, it must ensure that it protects American soldiers to the highest degree possible and improves its offensive capacities without inadvertently creating technologies that conflict with its other responsibilities.

To a large extent, this will be a matter of effective public diplomacy. The Army is sensitive to the demands of engaging with civilian audiences and shaping their attitudes about the Army and its mission.[49] A central part of this public diplomacy is the display of new weapons technologies and efforts to attract recruits who can operate them. The risk of inadvertently lowering barriers against war is not only linked to drones themselves but also to how drones are perceived. That is to say, the risk critics of drone warfare call attention to is the danger that a lack of American casualties will be confused with a sanitization of war even as wars continue to inflict terrible human suffering. The Army has an important role to play in presenting information about drones and informing the public about their associated risks and benefits. In particular, the Army should help to ensure that the American public does not lose sight of the fact that, despite the benefits drones bring in terms of force protection and offensive power, wars remain extremely destructive activities that must be waged for the right reasons and only as a last resort.

Policymakers face a different set of ethical challenges when it comes to developing and using drones. Although drones are unlikely to produce the kind of profound civic disengagement in military decisions that critics of drone warfare fear, these weapons nevertheless exert some influence on support for war, which could help policymakers build the case for war and escape a public backlash against costly military operations. Drones will be particularly important when they are used in conjunction with other strategies for justifying a war—for example, if a prospective military venture can be framed as a counterterrorism operation. This should not be considered a purely good or bad outcome. In some instances, war may be warranted and helping politicians make the case for war will be morally advantageous. At other times, a war may fail to pass *jus ad bellum* standards and any effect drones have in lowering inhibitions against fighting will be morally harmful.

Decisions about whether a war should be waged must be made on a case-by-case basis, which makes it impossible to pass final judgment on drones as being purely good or bad weapons. However, our results suggest that drones make other *jus ad bellum* considerations more important than ever. Differences in the principal policy objectives in our experiment showed that the perceived legitimacy and urgency of military intervention affected support for initiating hostilities. Unlike drones, which can lower inhibitions against fighting in just or unjust wars, principal policy objectives are central to determining whether war is justified. This makes it vital for policymakers to formulate just principal policy objectives and to state openly the principal policy objectives they are pursuing so these can be evaluated by the American public. It is critical to note that drawing out the implications of our findings for the military and policymakers will also depend on further research about how drones affect public opinion. The concern over drones evading the restrictive effects of casualty aversion can be extended to foreign civilians. Foreign civilian casualties may erode support for war just as military casualties do.[50] Members of the public may lose interest in wars that appear to be misdirected at innocent people or that inflict disproportionate civilian "collateral damage." This could give states an incentive to fight in ways that minimize the risk to foreign civilians just as they have an incentive to minimize military casualties. However, in this context the concern expressed by critics of drone warfare is not that drones will actually lower civilian casualty rates, as they may for military personnel, but that drones will give the appearance of reducing civilian victimization even as the increased incidence of war and new methods of fighting put civilians at greater risk than ever.

Kaag and Kreps[51] raise this challenge by suggesting that drones' impressive technical capacities may give the misleading impression that wars are being waged with greater attention to the *jus in bello* principles of proportionality and discrimination. This could, in turn, affect citizens' judgments relating to *jus ad bellum*. After all, if drones appear to improve compliance with the *in bello* rules of war, then they could also make the decision to wage war seem less morally significant. Drones might even give the appearance that the only casualties of war are enemy combatants. Closely related to this is the suspicion that the U.S. Government may already be underreporting the civilian casualties inflicted by drone strikes in an effort to make these strikes appear to be an attractive alternative to other types of military operations.[52] By this account, drones may be ethically objectionable because they facilitate dissimulation more than other weapons.

Several studies found that public opinion does not seem to be strongly affected by the suffering of foreign civilian populations.[53] There is evidence to show that democracies can inflict heavy foreign civilian casualties, and even target civilians, without sustaining any serious crises of public confidence.[54] This research would suggest that drones' capacities for creating the appearance of minimal civilian harm are relatively unimportant. After all, if public opinion is not sensitive to civilian casualties, then there is little reason to think that drones may reduce inhibitions against fighting by hiding those casualties.

Research that is focused specifically on drone strikes that harm civilians suggest that the apparent insensitivity to foreign civilian casualties has been overstated. Kreps[55] argues that the extent to which civilian casualties undermine support for drone strikes has been underestimated in polls and that it is stronger than the available data would suggest. She substantiates this with experiments that show how different ways of framing polling questions may elicit greater sensitivity to civilian casualties. Walsh[56] finds that the anticipated number of civilian casualties has a powerful influence on attitudes toward drone strikes—even exceeding the intolerance for sustaining military ca- sualties—and that sensitivity to civilian casualties appears to be higher when they are inflicted using precision weapons. This is evidence that the use of drones and other precision weapons may prime people to expect lower civilian casualty rates and cause them to adjust their willingness to condone them.

End Notes

[1] Medea Benjamin, *Drone Warfare: Killing by Remote Control*, New York: OR Books, 2012; Sarah Kreps and John Kaag, "The Use of Unmanned Aerial Vehicles in Asymmetric Conflict: Legal and Moral Implications," *Polity*, Vol. 44, No. 2, 2012, pp. 260-285; John Kaag and Sarah Kreps, *Drone Warfare*, Malden, MA: Polity Press, 2014.

[2] Robert Sparrow, "Killer Robots," *Journal of Applied Philosophy*, Vol. 24, No. 1, 2007, pp. 62- 77; Ronald C. Arkin, *Governing Lethal Behavior in Autonomous Robots*, Boca Raton, FL: Taylor & Francis Group, 2009; Armin Krishnan, *Killer Robots: Legality and Ethicality of Autonomous Weapons*, Burlington, VT: Ashgate, 2009; Aaron Johnson and M., Sidney Axinn, "The Morality of Autonomous Robots," *Journal of Military Ethics*, Vol. 12, No. 2,

2013, pp. 129-141; Marcus Schulzke, "Autonomous Weapons and Distributed Responsibility," *Philosophy and Technology*, Vol. 26, No. 2, 2013, pp. 203-219.

[3] John E. Mueller, *War, Presidents and Public Opinion*, New York: Wiley, 1973; Edward N. Luttwak, "Towards Post-Heroic Warfare." *Foreign Affairs*, Vol. 74, No. 3, 1995, pp. 109–122.

[4] Bruce Bueno de Mesquita and Randolph M. Siverson, "War and the Survival of Political Leaders: A Comparative Study of Regime Types and Political Accountability," *The American Political Science Review*, Vol. 89, No. 4, 1995, pp. 841-855. Louis Klarevas, "Trends: The United States Peace Operation in Somalia," *Public Opinion Quarterly*, Vol. 64, No. 4, 2000, pp. 523–540; Louis Klarevas, "The 'Essential Domino' of Military Operations: American Public Opinion and the Use of Force," *International Studies Perspectives*, Vol. 3, No. 4, 2002, pp. 417-437; Scott Sigmund Gartner, "The Multiple Effects of Casualties on Public Support for War: An Experimental Approach," *American Political Science Review*, Vol. 102, No. 1, 2008, pp. 95-106; M. A. Baum, and T. Groeling, "Reality Asserts Itself: Public Opinion on Iraq and the Elasticity of Reality, *International Organization*, Vol. 64, 2010, pp. 443-479; Douglas Kriner and Francis Shen, "Responding to War on Capitol Hill: Battlefield Casualties, Congressional Response, and Public Support for the War in Iraq," *American Journal of Political Science*, Vol. 58, No. 1, 2014, pp. 157-174; H. E. Goemans, "Fighting for Survival: The Fate of Leaders and the Duration of War," *The Journal of Conflict Resolution* Vol. 44, No. 5, 2000, pp. 555-579; David L. Rousseau, Trevor Thrall, Marcus Schulzke, and Steve S. Sin, "Democratic Leaders and War: Simultaneously Managing External Conflicts and Domestic Politics," *Australian Journal of International Affairs*, Vol. 66, No. 3, 2012, pp. 349-364.

[5] Christopher Gelpi, Peter D. Feaver, and Jason Reifler, "Success Matters: Casualty Sensitivity and the War in Iraq," *International Security*, Vol. 30, No. 3, 2006, pp. 7-46; Christopher Gelpi, Peter D. Feaver, and Jason Reifler, *Paying the Human Costs of War: American Public Opinion & Casualties in Military Conflicts*, Princeton, NJ: Princeton University Press, 2009.

[6] Gelpi, Feaver, and Reifler, *Paying the Human Costs of War*, p. 2.

[7] Eric V. Larson, *Casualties and Consensus: The Historical Role of Casualties in Domestic Support for U.S. Military Operations*, Santa Monica, CA: RAND, 1996, p. 101.

[8] Charles K. Hyde, *Casualty Aversion: Implications for Policy Makers and Senior Military Officers*, Newport, RI: Naval War College, 2000, p. 10.

[9] Cori Dauber, "Image as Argument: The Impact of Mogadishu on U.S. Miltiary Intervention," *Armed Forces & Society*, Vol. 27, No. 2, 2001, pp. 205-230; Cori Elizabeth Dauber, "The Shot Seen 'Round the World': The Impact of the Images of Mogadishu on American Military Operations," *Rhetoric & Public Affairs*, Vol. 4, No. 4, 2001, pp. 653-687.

[10] Douglas Kriner and Francis Shen, "Responding to War on Capitol Hill: Battlefield Casualties, Congressional Response, and Public Support for the War in Iraq," *American Journal of Political Science*, Vol. 58, No. 1, 2014, pp. 157-174.

[11] Charles K. Hyde, *Casualty Aversion: Implications for Policy Makers and Senior Military Officers*, Newport, RI: Naval War College, 2000.

[12] Kaag and Kreps, *Drone Warfare*, p. 2.

[13] *Ibid.*, p. 65.

[14] *Ibid.*, p. 76.

[15] P. W. Singer, *Wired for War: The Robotics Revolution and Conflict in the 21st Century*, New York, Penguin Press, 2009, p. 255.

[16] *Ibid.*, p. 258.

[17] Frank Sauer and Niklas Schörnig "Killer Drones: The 'Silver Bullet' of Democratic Warfare?" *Security Dialogue*, Vol. 43, No. 4, 2012, pp. 363-380.

[18] Daniel Brunstetter and Megan Braun, "The Implications of Drones on the Just War Tradition," *Ethics & International Affairs*, Vol. 25, No. 3, 2011, pp. 337-358.

[19] *Ibid.*, p. 346.

[20] Linda Johansson, "Is It Morally Right to Use Unmanned Aerial Vehicles (UAVs) in War?" *Philosophy & Technology*, Vol. 24, No. 3, 2011, p. 283.

[21] Christian Enemark, *Armed Drones and the Ethics of War: Military Virtue in a Post-Heroic Age*, New York: Routledge, 2013.

[22] *Ibid.*, p. 23.

[23] *Ibid*; Christian Enemark, "Drones over Pakistan: Secrecy, Ethics, and Counterinsurgency," *Asian Security*, Vol. 7, No. 3, 2011, pp. 218-237.

[24] David Hastings Dunn, "Drones: Disembodied Aerial Warfare and the Unarticulated Threat," *International Affairs*, Vol. 89, No. 5, 2013, pp. 1237-1246.

[25] *Ibid.*, p. 1238.

[26] Michael J. Boyle, "The Costs and Consequences of Drone Warfare," *International Affairs*, Vol. 89, No. 1, 2013, pp. 1-29.

[27] Metin Gurcan, "Drone Warfare and Contemporary Strategy Making: Does the Tail Wag the Dog?" *Dynamics of Asymmetric Conflict: Pathways Toward Terrorism and Genocide*, Vol. 6, No. 1-3, 2013, pp. 153-167.

[28] Zack Beauchamp and Julian Savulescu, "Robot Guardians: Teleoperated Combat Vehicles in Humanitarian Military Intervention," Bradley Jay Strawser, ed., *Killing by Remote Control: The Ethics of an Unmanned Military*, New York: Oxford University Press, 2013.

[29] Bradley Jay Strawser, "Moral Predators: The Duty to Employ Uninhabited Aerial Vehicles," *Journal of Military Ethics*, Vol. 9, No. 4, 2010, pp. 342-368.

[30] Ronald C. Arkin, *Governing Lethal Behavior in Autonomous Robots*, Boca Raton, FL: Taylor & Francis Group, 2009; Strawser, "Moral Predators," pp. 342-368; Marcus Schulzke, "Robots as Weapons in Just Wars," *Philosophy and Technology*, Vol. 24, No. 3, 2011, pp. 293-306; Marcus Schulzke, "The Morality of Remote Warfare: Against the Asymmetry Objection to Remote Weaponry," *Political Studies*, 2014; Anton Petrenko, "Between Berserks-gang and the Autonomous Weapons Systems," *Public Affairs Quarterly*, Vol. 26, No. 2, 2012, pp. 81-102.

[31] Michael Walzer, *Just and Unjust Wars: A Moral Argument with Historical Illustrations*, New York: Basic Books, 1977, pp. 155- 157; Jeff McMahan, "The Just Distribution of Harm between Combatants and Noncombatants," *Philosophy & Public Affairs*, Vol. 38, No. 4, 2010, pp. 342-379.

[32] Adam J. Berinsky, Gregory A. Huber, and Gabriel S. Lenz, "Evaluating Online Labor Markets for Experimental Research: Amazon. com's Mechanical Turk," *Political Analysis*, Vol. 20, 2012, pp. 351–368; T. S. Behrend, D. J. Sharek, A. W. Meade, and E. N. Wiebe, "The viability of crowdsourcing for survey research," *Behavior Research Methods*, Vol. 43, No. 3, 2011, pp. 800–813.

[33] A total of 3,784 individuals participated in the experiment. Each participant was paid $.40. The survey was conducted in January 2015. The resulting data, and code used to generate the results reported here are available from *www.jamesigoewalsh.com*.

[34] B. J. Gaines, J. H. Kuklinski, and P. J. Quirk, "The Logic of the Survey Experiment Reexamined," *Political Analysis*, Vol. 15 No. 1, 2007, pp. 1–20.

[35] Bruce W. Jentleson, "The Pretty Prudent Public: Post Post-Vietnam American Opinion on the Use of Military Force," *International Studies Quarterly*, Vol. 36, pp. 49-74; Gelpi, Feaver, and Reifler, *Paying the Human Costs of War.*

[36] Gelpi, Feaver, and Reifler, *Paying the Human Costs of War.*

[37] Full results of these models are reported in the online appendix available from *www.jamesigoewalsh.com*. The estimation technique is ordinary least squares regression. Ordered logistic regression produces substantively similar results. See the online data and code for details. We also included two measures of general attitudes towards the use of force as independent variables—militant assertiveness and isolationism (drawn from Richard K. Herrmann, Philip E. Tetlock, and Penny S. Visser, "Mass Public Decisions on Go to War: A Cognitive-Interactionist Framework," *American Political Science ReviewI*, Vol. 93 No. 3, 1999, pp. 553-573). Including these independent variables did not alter the substantive size or statistical significance of the variables for drone strikes.

[38] Pamela Johnston Conover and Virginia Sapiro. "Gender, Feminist Consciousness, and War," *American Journal of Political Science*, Vol. 37, 1993, pp. 1079-1099.

[39] Note that in our experiment, gender does not influence attitudes towards the use of force for humanitarian intervention (see Figure 8). This is consistent with earlier works which find that women are as or more likely to support military action in such cases. See Deborah Jordan Brooks and Benjamin A. Valentino, "A War of One's Own: Understanding the Gender Gap in Support for War," *Public Opinion Quarterly*, Vol. 75, No. 2, 2011, pp. 270-286.

[40] E.g., Peter Hays Gries, *The Politics of American Foreign Policy: How Ideology Divides Liberals and Conservatives Over Foreign Affairs*, Stanford, CA: Stanford University Press, 2014; Richard Johnston, "Party Identification: Unmoved Mover or Sum of Preferences?" *Annual Review of Political Science*, Vol. 9, 2006, pp. 329-351.

[41] Bueno de Mesquita and Siverson, "War and the Survival of Political Leaders"; Goemans, "Fighting for Survival"; Rousseau, Thrall, Schulzke, and Sin, "Democratic Leaders and War."

[42] Sauer and Schörnig "Killer Drones: The 'Silver Bullet' of Democratic Warfare?"

[43] Singer, *Wired for War.*

[44] Kreps and Kaag, "The Use of Unmanned Aerial Vehicles in Asymmetric Conflict"; Kaag and Kreps, *Drone Warfare.*

[45] Gelpi, Feaver, and Reifler, *Paying the Human Costs of War.*

[46] Jentleson, p. 64.

[47] Beauchamp and Savulescu, "Robot Guardians."

[48] Paul Robinson, "Ethics Training and Development in the Military," *Parameters,* Spring 2007; Chris Case, Bob Underwood, and Sean T. Hannah, "Owning Our Army Ethic," *Military Review: The Army Ethic*, 2010, pp. 3-10; Jeffrey Wilson, "An Ethics Curriculum for an Evolving Army," Paul Robinson, Nigel De Lee, and Don Carrick, eds., *Ethics Education in the Military*, Burlington, VT: Ashgate Publishing, Ltd, 2008.

[49] Michael Pasquarett, ed., *Perspectives on Embedded Media: Selected Papers from the U.S. Army War College*, Carlisle, PA: US Army War College, 2004; Major Edward L. English, *Towards a More Productive Military-Media Relationship*, Fort Leavenworth, KS: School of Advanced Military Studies, U.S. Army Command and General Staff College, 2005; Major Chad G. Carroll, "The U.S. Army Public Diplomacy Officer: Military Public Affairs Officers" Roles in the Global Information Environment," *Master's Thesis*, Chapel Hill, NC: University of North Carolina at Chapel Hill, 2007.

50 Richard C. Eichenberg, "Victory Has Many Friends: U.S. Public Opinion and the Use of Force," *International Security*, Vol. 30, No. 1, 2005, pp. 140–177; Sarah Kreps, "Flying under the Radar: A Study of Public Attitudes Towards Unmanned Aerial Vehicles," *Research & Politics* 2014, available from *rap.sagepub.com/content/1/1/2053168014536533*; James Walsh, "Precision Weapons, Civilian Casualties, and Support for the Use of Force," *Political Psychology*, 2015.

51 Kreps and Kaag, "The Use of Unmanned Aerial Vehicles in Asymmetric Conflict"; Kaag and Kreps, *Drone Warfare.*

52 Trevor McCrisken, "Obama's Drone War," *Survival: Global Politics and Strategy,* Vol. 55, No. 2, 2013, pp. 97-122.

53 John Mueller, "Public Opinion as a Constraint on U.S. Foreign Policy: Assessing the Perceived Value of American and Foreign Lives," National Convention of the International Studies Association, 2000; John Tirman, *The Deaths of Others: The Fate of Civilians in America's Wars*, New York: Oxford University Press, 2011.

54 Alexander B. Downes, "Desperate Times, Desperate Measures: The Causes of Civilian Victimization in War," *International Security*, Vol. 30, No. 4, 2006, pp. 152-195; Alexander B. Downes, "Restraint or Propellant? Democracy and Civilian Fatalities in Interstate Wars," *The Journal of Conflict Resolution*, Vol. 51, No. 6, 2007, pp. 872-904; Alexander B. Downes, *Targeting Civilians in War*, Ithaca, NY: Cornell University Press, 2008.

55 Kreps, "Flying under the Radar."

56 Walsh, "Precision Weapons, Civilian Casualties, and Support for the Use of Force."

ABOUT THE AUTHORS

MARCUS SCHULZKE is a Post-Doctoral Research Fellow in the School of Politics and International Studies at the University of Leeds, West Yorkshire, England. He has published research on the morality of drone warfare, moral decisionmaking in combat, and just war theory for such journals as *Political Studies, The Journal of Military Ethics*, and *Philosophy and Technology*. Dr. Schulzke holds a Ph.D. in political science from the University at Albany, NY, with a dissertation on how soldiers make ethical decisions during counterinsurgency operations.

JAMES IGOE WALSH is Professor of Political Science at the University of North Carolina at Charlotte, NC. His research interests include the military and political consequences of advanced weapons, links between natural resources and conflict, and intelligence and national security. His work has been supported by the Army Corps of Engineers, the Department of Homeland Security, the National Science Foundation, and the Minerva Research Initiative. Dr. Walsh is the author of *The International Politics of Intelligence Sharing*, published by Columbia University Press, New York, and was named an Outstanding Title by Choice. Dr. Walsh holds a Ph.D. in international relations from American University.

INDEX